Clottie & Bob
Merry Christmas - 1986
Bob

BIRDS AND MAMMALS

NATIONAL WILDLIFE FEDERATION

Library of Congress CIP Data: page 254

BIRDS AND MAMMALS

National Wildlife Federation's
50th Anniversary Collection of Favorites
from National *and* International Wildlife *Magazines*

INTRODUCTION

A great nature writer must be brimming over with curiosity and really turned on by the outdoors. A top wildlife photographer must have creative skills, endless patience, and a willingness sometimes to take risks that would scare the stuffing out of you and me. The wildlife scientist has all of these qualities in good measure.

This titillating combination has made my job of editing *National Wildlife* and *International Wildlife* magazines the last 25 years a joy and an adventure. Every issue! Because, through their eyes and art and dedication, we can bring you a close-up of nature and all of its mystery, beauty, and wonder.

So, for a very special occasion—the 50th Anniversary of the National Wildlife Federation—we have compiled this treasury of "Birds and Mammals" from articles that have appeared in our magazines. It is both a tribute to those extraordinary people who help us see nature through the word and lens, and a celebration of nature itself. Let me give you a taste of what you can expect to find inside this special, golden anniversary collection.

Just imagine being trapped on a large ice floe in the Bering Sea, unable to get back to shore, and seeing a gigantic polar bear ambling toward you across the packed ice. Writer-photographer Fred Bruemmer found himself in this predicament, but, with the help of Eskimos, got out to tell us his story.

Or consider the dedication of biologist Pete Myers. For the past 10 years, he has followed an amazing eight-inch puff of feathers known as the sanderling from North America to South America—and back. This little bird flies 8,000 miles a year, sometimes 35 hours without stopping.

You may think of the deer as a gentle, elusive animal. But George Harrison will tell you how nature writers often flirt with unknown dangers—such as being attacked by a 350-pound whitetail buck during the rutting season.

After all these years, I'm still continually surprised at the wonderful stories and pictures our contributors bring to *National Wildlife* and *International Wildlife* magazines. But maybe I shouldn't be. Wildlife seems to bring out the dedication in people who work with it—a prime reason why the National Wildlife Federation has been America's number one advocate for nature for 50 years.

John Strohm

EDITOR, *National Wildlife* and *International Wildlife* magazines

BIRDS

Feathers

STORY AND PHOTOGRAPHS BY
DAVID CAVAGNARO

I sat for a long time on the west cliffs of Isla San Martín in Baja California, watching the slow steady progress of gray whales heading south. At my feet, young grasses and succulent leaves of ice plant glistened with droplets of water. As my eyes explored the wet carpet, a little feather, white against the green, caught my attention. Bending down close, I could see, as through a lens, the intricate structure of the feather, magnified by the beads of rain left delicately balanced there.

The feather was from the breast of a pelican, and was apparently still as good as new after a year's service at sea. Through the magnifying drops of rain I could see the filaments, or barbs, that make up the vanes on either side of the quill. Where the barbs had separated, I could even see the fine barbules that line the sides of each barb. The barbules are equipped with microscopic hooks which lock into each other in much the same way as the two sides of a zipper fit together. It is this remarkable piece of engineering that allows a bird to return its feathers quickly to functional condition after rough treatment. With a little oiling and preening, each feather is zipped back together again, its ability to insulate and waterproof fully restored.

In flight, a bird's life depends upon its feathers. Large birds may have as many as 25,000, each of which contributes to the streamlining of its body and each of which must from time to time be preened and kept in good condition.

But none are more important in the air than the strong, durable flight feathers of the wings and tail. One large, primary flight feather may consist of a thousand barbs and a million barbules. Each tiny component within a feather, each feather on a wing, each bone and muscle that lies beneath, all work together in an aerodynamic design, the efficiency and versatility of which are still the envy of the best aeronautical engineers.

The feather itself is at least 140 million years old. The earliest known fossil feather, perfectly preserved in fine-grained limestone, was found in a Bavarian quarry where lithographic stones were being carefully mined. The fossil was identical in every apparent respect to a modern-day feather. Significantly, the vane on one side of the quill was wider than the vane on the other side. The asymmetry is a feature of flight feathers only and would eventually be recognized as certain proof that the owner of that ancient feather could fly. Soon afterwards a complete skeleton of this first feathered lizard burst upon the scientific world: *Archaeopteryx lithographica,* "ancient wing of lithographic stone."

Today, birds are one of the largest classes of vertebrates, exceeded only by fish. They comprise 8,700 species ranging in size from the condor, with its wingspan of ten feet, to the tiny hummingbird. Meanwhile mammals, which evolved on a nearly identical time schedule, have given us only 4,000 present-day species. Throughout these millions of years of change, dramatic enough for birds of flight and mammals of the land to return to the sea as penguins and whales, the basic types of feathers once worn by the primordial lizard bird have emerged essentially unchanged.

Feathers serve two primary functions: thermoregulation and flight. The feathers of totally flightless birds, however, like the kiwi, ostrich, cassowary, and emu, are used mainly for thermoregulation and so have lost, for lack of need, the hooked barbules that make normal feathers aerodynamically sound. Their resulting body covering looks and functions remarkably like the fur of mammals. Highly specialized divers such as the penguins, which also no longer fly, have developed a dense covering of hairlike body feathers, which like the fur coat of seals is well adapted to the frigid waters of Antarctica.

The origin of the feather, like God and the beginning of the universe, still eludes us. Whenever I see a feather on the ground, like the one in Baja covered with drops of rain, I feel privileged to stand for just a moment in the presence of a great question. That feeling is perhaps as satisfying as discovering the answer itself. (continued)

David Cavagnaro is a California-based writer and photographer, whose most recent book is Feathers *(Graphic Arts Center Publishing Company, Portland, Oregon, 1982), from which this text and photography was adapted.*

*B*rilliant back feathers on a male
golden pheasant not only attract mates
but also help regulate body temperature.

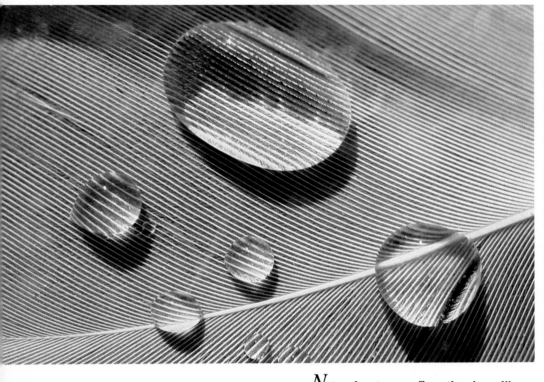

*N*atural waterproofing, the zipperlike design of a pelican feather (above) keeps raindrops from penetrating to the body.

*B*acklighting by the sun creates a prismatic effect, turning brown varied thrush wing feathers (right) to rainbows.

*H*ypnotic "eyes" on a male peacock's nuptial plumage intensify the creature's courtship display during breeding season.

Fairy Tales

. . . At least when it comes to the fairy tern, which performs one of nature's great balancing acts

Are True

BY LISA YOUNT
PHOTOGRAPHS BY FRANS LANTING

I	T IS ONE of the bird world's finest balancing acts: while other species are busy building nests, the fairy tern simply lays its single egg in the open atop a narrow branch. Then, with the help of its mate, the creature holds the egg in place for a month until the chick finally emerges. The act is performed even when there are no trees. Throughout Hawaii and the other Pacific islands where they range, the snow-white seabirds can be seen balancing their eggs on rocky ledges, woodpiles, rooftops, even windowsills.

Called *Manu O Ku* by early Hawaiians, the fairy tern is one of six species of terns that regularly raise their young in the fiftieth state, and one of some 40 species of terns that live throughout the world. Its plumage is pure white, except for rings of black feathers around its dark blue eyes. Seen against the sun, the bird appears almost translucent. It is fairly small compared to many other members of its family, with a 10-inch body and a 28-inch wingspan. Its tail is slightly forked—a trait that caused naturalists to nickname some terns "sea swallows." (The name "tern" itself derives from the sound of the piercing *tee-ar-r-r*

Illuminated against the Hawaiian sky, a fairy tern displays the translucent, ethereal beauty that has earned it a second name: "the Holy Ghost bird."

On Pacific Islands, young terns hatch in the most unlikely places

A fairy tern needs a delicate sense of balance from the moment it emerges, as this hatching chick is demonstrating (left). Entering the world atop a water faucet in the Northwestern Hawaiian Islands (bottom left), the tiny bird is already well developed with long claws that can grasp the precarious perch its parents have chosen. Adult fairy terns build no nests; instead, they lay their eggs on almost any available surface. This youngster, standing on the branch where it was hatched (right), will stay with its parents for two months.

cries that many members of the family scream.)

At one time or another, the fairy tern has been known as the white tern, the love tern (because of the mated pair's habit of stroking and preening each other's faces) and the "Holy Ghost bird" (because of the creature's ethereal appearance in the air). Like many other seabirds, it apparently cannot recognize its own egg or chick. Scientists have found that the bird will adopt foster chicks, and it will also incubate another tern's egg. In one experiment in the 1950s, researcher Alfred M. Bailey substituted a piece of rounded coral for a fairy tern egg, and then watched an adult return and settle down to "incubate" the coral.

Among terns, fairies are not the only seemingly "careless" parents. Several other species lay their eggs directly on the ground, after making only a few scrapes with their claws to prepare the site. Depending on the type of tern, the eggs may be laid on the seashore, in a salt or freshwater marsh, or on the edge of a pond. Least terns, a species commonly found in New England, sometimes lay their eggs so close to the water that the high tide sweeps them away.

This apparent carelessness is deceptive, however. The speckled eggs of the ground-nesting terns are all but invisible on the mottled sand of a beach, making them less vulnerable to predators. Fairy terns are also more careful than they might appear. They settle cautiously over their precariously

perched eggs and almost never knock them down.

Male and female fairy terns take turns incubating the egg continuously during the 35 days before it hatches. In the creature's tropical habitat, this incubation is as necessary to protect the egg from heat as from cold. Fairy tern parents also take turns caring for and feeding their chick for at least two months after its birth.

The chick is born well developed, with long sharp claws that help it to hold onto whatever strange perch its parents have chosen. On Midway Island, University of Hawaii biologist Andrew J. Berger found adults brooding chicks at the very edge of flat-roofed buildings. He also saw one bird incubating an egg on the stub of an ironwood tree that had been broken off about 30 feet above the ground.

Like any other nestling, the young fairy tern is a ball of fluff and has an inexhaustible appetite. Both parents are kept busy supplying it with fish and squid. The adult terns hover over the water and dive down to catch any prey jumping at the surface. What's more, the birds can continue foraging while holding onto other fish already caught. A fairy tern may come back to its chick with up to a dozen fish laid crosswise in its bill. The chick usually will swallow them whole, one by one.

Compared to some other members of the tern family, the fairy tern is a homebody. It rarely ventures far from its breeding islands. Many other terns,

On Midway Island in the Pacific, an adult incubates its speckled egg on an ironwood tree branch. The two parents take turns warming the egg during the month prior to hatching.

however, are migrants. The common terns of New England may fly as far south as Rio de Janeiro in Brazil, for the winter. Roseate terns, also found in the New England states, sometimes migrate to Guyana in northeastern South America. The arctic tern makes perhaps the most spectacular journey of all. It travels yearly from the Arctic to Antarctica and back—a trip of up to 12,000 miles each way.

To people who have studied the bird, the fairy tern's inquisitiveness is often as striking as its strange nesting behavior. A few years ago, Ralph Schreiber, curator of ornithology at the Natural History Museum in Los Angeles, found himself surrounded by a hovering cloud of the birds on Christmas Island in the South Pacific. "They hover over your head at just the right angle," says the scientist. "That's why you see so many photos of them silhouetted against the sun."

On Christmas Island, Schreiber found one fairy tern that had deposited its egg on a wheel-like faucet attached to a garden hose. Taking turns with its mate, the bird steadfastly balanced the egg in place until the chick hatched. That, indeed, is a tough act to follow. ▣

California writer Lisa Yount believes that one fairy tern deserves another. Another Californian, Frans Lanting, photographed the birds pictured here in the Northwestern Hawaiian islands.

The fairy tern is pure white, except for a narrow black ring around its eyes (opposite page). Its predilection for unusual egg-laying sites includes a pile of driftwood on a remote island beach.

PHOTOGRAPHS BY JEN AND DES BARTLETT

Diary of a loon watcher

from the journal of Jen and Des Bartlett

We stayed inside the blind for hours, emerging stiff and cold after being attacked by hordes of mosquitoes

WITH its dramatic markings and imposing size, the loon is a wildlife photographer's delight. Unfortunately, though, this magnificent symbol of the Canadian North is difficult to approach, and anyone who does get close must contend with a few inconveniences: mosquitoes, rain squalls, boggy ground. For two summers, being careful not to harass the birds, we put up with these and other distractions in remote parts of the Northwest Territories. This photo collection is one result of our persistence.

Two of the world's four loon species breed during the short summer in the tundra where we camped. The most difficult to photograph was the Arctic loon, which prefers nesting on small islands in large tundra lakes. The slightly smaller red-throated loon was easier to work with because it chooses locations on banks of smaller lakes or streams.

To get close to the red-throated loons, we used eight separate blinds. These small, canvas tents withstood all but the strongest winds, but when the permafrost turned to mush later in the summer, tent pegs were useless, and we secured the corners with large rocks. During severe storms, wind gusts reached 45 miles an hour or more. Then, all we could do was clutch the door and window flaps to secure them against the wind and horizontally driven rain.

Photographing the Arctic loon required even more engineering. To get close to one nest, we constructed an artificial island of rocks and built a plywood box on it for a blind. This nest was in a lake near camp, but getting out to it with our equipment was an ordeal. The lake was two feet deep with an uneven spongy bottom, and one slip would have put several thousand dollars worth of equipment into the drink. We remained inside the blind for hours at a time, emerging stiff and cold after being attacked by hordes of mosquitoes. But our patience was rewarded when the first egg hatched in early August.

AT the nest, the loon rolls the blotched, brown eggs with its bill until it is satisfied with the arrangement. Then, the bird settles down to incubate, a four-week-long task shared by both parents. One chick usually hatches a day or two before the second appears.

At two red-throated loon nests we watched, the eggs hatched in early August. During the first days of the chicks' lives, one adult remained with them while the other flew off. Returning about an hour later with a fish in its beak, the parent swam toward the brooding adult. One or both chicks slipped into the water to take the fish. But what a fish! Often, it was five inches long. Occasionally, the chicks refused such large meals. When a youngster did attempt to gulp a fish down, however, a battle was on. Unwilling to lose its catch should the chick drop it, the adult held on while the youngster struggled to swallow it headfirst. Burdened with the weight of the fish, the chick kicked its tiny feet frantically to keep its own head above water. Once the little loon succeeded, the fish tail would sometimes protrude from its beak for an hour or so.

Several times we watched as a strange loon landed. Each time the resident pair rushed across the water at the interloper, standing almost erect on the surface with bodies out of the water. Sometimes they flew for some distance to chase the intruder well away. Loons are strong fliers, but before they take off, they must flap and patter along the surface for a considerable distance.

For all the excitement, our look at the youngsters was short-lived. By the time their second chicks were 48 hours old, the loon families no longer used their nests as a permanent base. From then on, the best we could do was check their progress through binoculars.

Our patience was rewarded with the chance to watch the private lives of loons up close on their breeding ground

Its bright, red eye ever on the lookout for danger, the loon makes itself almost invisible to predators

AT the first sign of any predator, a nesting loon flattens its head and neck against the ground, making itself almost invisible from a distance. If the danger increases, the bird slides on its belly down a short, muddy runway into the water, submerging silently and swimming away underwater to surface far from the nest. A loon has almost the same specific gravity as water and can submerge with scarcely a ripple.

Jaegers and herring gulls commonly prey on the eggs and young of tundra-nesting birds, as do Arctic foxes and short-tailed weasels. Even so, only one loon pair we observed lost its eggs to predators. The nests are little more than scrapes on the damp soil, but the precautions taken by the parents to hide the locations are effective.

The bird is so skittish that even harmless sounds may make it cower. We watched one day as a loon flattened to the ground in response to the distant call of a sandhill crane. But sometimes the loons were aggressive themselves. One Arctic loon nest was only 20 feet from a whistling swan nest. The two species seemed compatible neighbors, but one day a loon attacked a male swan which appeared to be preening innocently on a nearby rock.

Occasionally, the birds resort to a bag of tricks when worried. As we waded to our main Arctic loon blind, the sitting adult would slip unseen into the lake. Sometimes as we neared the nest, the loon came closer and pretended to be injured, flapping and splashing on the surface then diving noisily as it tried to lure us away. But we performed a little magic of our own. An assistant always helped us carry our equipment to the blind. When he walked off, the loons were fooled into thinking we had left, too.

From our hidden vantage point, we could watch the youngsters snuggle under the parent's wings, their heads sometimes poking out. That was ordinary enough behavior, but from time to time we saw some puzzling activities, too. Once, an adult grasped a broken egg shell in its beak, swam a few feet from the nest, then let the shell sink to the bottom. Apparently, it was hiding a telltale sign that might have attracted an enemy to the nest — still another precaution against tundra predators. ◼

Des Bartlett, with his wife Jen and daughter Julie, has been filming wildlife since 1952. Camping for long periods close to their photographic subjects, they have shot nearly 3,000 films, including "The Land of the Loon," and exposed 1,372,600 feet of 16mm color film.

STORKS
get a helping hand

*Buddhist monks in Thailand are preserving a species,
but are they doing the job too well?*

By Nigel Sitwell
Photographs by Michael Freeman

THE OLD MONK beckoned me to follow. Dressed in bright saffron robes, he led me through the heavy wooden doors of Wat Phai Lom, an ancient Buddhist temple near Bangkok, Thailand. Within, he lit candles beneath the watchful gaze of gold-encrusted Buddhas and motioned for me to sit with him in the room's quiet shadow.

The calm of the temple's interior could not have been more at odds with its surroundings. Outside of the massive doors lives a riotous gathering—approximately 17,000 Asian openbill storks, the largest nesting colony of the snail-eating birds anywhere. The smell of ammonia from their droppings is sharp and all-pervasive, and the vociferous birds raise a continuous clamor. Their untidy nests are everywhere, stacked densely and sometimes precariously on every available perch.

"I was born near here," the monk told me, "and I remember a time nearly 60 years ago when there were

The abbot of Wat Phai Lom (above) aids an injured openbill stork. Thanks to the monks' respect for living things, their monastery now has 17,000 birds.

Bamboo is a key nesting material for the 31-inch-long birds (right), but supplies at the temple are dwindling.

only about 1,000 of the storks. In those days, people used to kill them with slingshots or guns. Boys collected their eggs to sell in the market as duck eggs, which they resemble. But this doesn't happen any more. The temple grounds are now a government wildlife sanctuary, and poaching has been stopped by police and the Royal Forest Department rangers stationed here."

The creation of the refuge is largely responsible for saving Thailand's sole remaining population of openbill storks. But it only reinforces a more basic protection the birds receive from the Buddhist monks. In its reverence for every living thing, Buddhism takes a passive and accepting attitude toward the environment. Reflecting this, the residents of Wat Phai Lom have allowed the storks to proliferate unchecked and undisturbed on the plot of holy land. Ironically, the birds may now be proliferating too well, for their excessive numbers are beginning to destroy the habitat they depend on. Some fear that the

monks' concern for individuals instead of the overall group may actually be putting the birds in jeopardy.

Wat Phai Lom ("Wat" means temple, and "Phai Lom" means surrounded by bamboo) is about an hour's drive north of Bangkok. Monks who live there depend on gifts of food donated by area residents. Since only a few families still live in the vicinity, Wat Phai Lom can support only five monks. Each day, many visitors come to view Thailand's only remaining colony of the storks. Few have given donations.

The sanctuary's success is rooted in Buddhist belief, the teachings of Siddhartha Gautama, who was born in what is now Nepal over 2,500 years ago. Siddhartha renounced worldly things at the age of 29 and embarked on a quest for knowledge. His search ended six years later when he attained enlightenment and became the Buddha, or the Enlightened One. From then until his death at 80, he devoted himself to teaching.

In some ways, Buddhism seems more a philosophy than a religion, for it recognizes no god and has no authority comparable to the Bible or the Koran. The tradition says that Buddha himself was neither a god nor a prophet, but a remarkable human being who proclaimed a way of achieving wisdom, compassion and freedom from suffering. Perhaps Buddhism can best be described as a way of life whose ultimate goal is the attainment of *nirvana,* a state of consciousness where all greed, hate and delusion have been extinguished—an end to suffering.

Buddhist thought holds that we all have lived many previous lives, and will live many more in the future; misdeeds in one life may have consequences in the next. But these lifetimes may not all be human. For this reason, Buddhism teaches concern for the sanctity of both human and animal life. For example, Sri Lanka's Buddhist King Upatissa I (368-410 A.D.) is said to have swept ants and insects out of his path with a broom made of peacock feathers so that he would not accidentally step on the creatures. And that nation's twelfth-century ruler, King Nissankamalla, proclaimed protection for all wildlife, including fish, within 25 miles of his city.

It is not unusual to find Buddhist shrines and temples serving as *de facto* animal sanctuaries. The Snake Temple in Penang, Malaysia, is well known for its pit vipers, which are most active at night, when eggs are left about for them to feed on. The vipers can also be seen during the day dozing on altars and curled up under tables.

Every monk vows on his ordination to abstain from destroying life, a precept also followed by many pious laypersons.

Within an ancient sanctuary, hidden from the clamor of nearby birds, a monk (above) sits beneath Buddha's gaze.

Standing room only! Now there are too many storks in the monastery for their own good. They strip nesting trees (left) and kill them with droppings.

But Buddhism is an extremely tolerant religion, and this tolerance is extended even to those who kill. A devout Buddhist might eat a chicken, for example, if it has been killed by someone else. Furthermore, if a wrong act—even including murder—is committed unwittingly, it carries no penalty.

Reverence for life, therefore, is not necessarily a guarantee that animals will not be harmed. In Thailand, there is a widespread practice of gaining merit by releasing small caged birds. But this has led to the absurd situation where a number of people find it profitable to catch such birds specifically to sell to those who wish to release them. Buddhist ethics are more concerned with passive protection than with active conservation or the survival of larger communities. Some Buddhists may feel no concern when, say, a forest is cut down, because they themselves are not doing the cutting, and because logging does not directly kill animals.

But logging deprives the openbill stork of needed nesting sites. The Asian openbill occurs across a broad swath from Pakistan to Indochina. It is closely related to the African openbill and allied to the wood stork of the southeastern United States. Clothed early in the breeding season in immaculate black and white plumage with bright pink legs, the adults assume a dirty gray color soon after their eggs are laid. The legs also become less bright.

Migrating openbills begin to arrive at Wat Phai Lom in October, most likely from Bangladesh. In November, nesting begins. The stork nests are close together, often less than two feet apart. The birds prefer large trees where their nests will be high above ground. Sugar palms are the first choice, followed by other large trees, and then by bamboo.

The normal clutch is four eggs. Chick mortality is high as many eggs fall to the ground, especially during storms. Generally only two survive. Incubation lasts about one month, and the young fledge in another 40-45 days. Activity in the colony peaks in February, when some 90 percent of the nests have young. In good years there may be two or even three shifts at a particular nest site, as the couples that want to breed take their turns.

During the day the adult storks make frequent journeys to their feeding grounds in the surrounding rice paddies. Favored food: large round apple snails. A feeding stork probes shallow water with its bill. Snails are carried to dry ground where the meat can be extracted. While steadying the snail's shell with its upper mandible, the stork uses its lower one to pry into the opening, severing the muscle that attaches the snail's body to its shell.

After World War II, Thailand's openbill storks fell victim to increased hunting and deforestation. By 1955 the colony at Wat Phai Lom was Thailand's last. In that year Thailand's leading conservationist—Dr. Boonsong Lekagul—heard that people were shooting storks near the temple. So he persuaded the monks to declare the area occupied by the storks to be part of the temple grounds, thus providing sanctuary. In 1960, the openbill stork was granted full legal protection in Thailand, and in 1970, the Wat Phai Lom site was given official status as a sanctuary.

The storks have responded well. In 1964 the number of birds was estimated at 4,000, but by 1978 the population had quadrupled. Yet, the future of the colony is not assured. One threat is posed by the wide use of pesticides in the neighboring rice fields. These could accumulate in the apple snails and be passed to the storks.

Another potential threat lies in the storks' behavior. Each morning, the adults take to the air and gain altitude from thermal air currents before heading for their feeding grounds. Unfortunately, Wat Phai Lom lies close to the flight path for aircraft landing at Bangkok's Don Muang Airport, and the storks' morning flights coincide with the peak period of European airline arrivals.

But the main problem facing the stork colony arises from the huge volume of excrement produced by the birds themselves. The accumulation is gradually killing the important nesting trees. Many of the bigger trees, especially the huge Asian hardwoods known as *dipterocarps,* have already vanished. In times past, as many as 386 stork nests could be found in a single tree.

Formerly too, annual floods spilled from the river to inundate the sanctuary. "This was effective in washing the soil," explains Virach Chantrasmi, president of the Bangkok Bird Club. "But recently," he adds, "dams have been built and the Chao Phya River seldom breaks its banks, so the trees have no respite from the droppings."

As the big trees have disappeared, the storks have become much more dependent on the clumps of bamboo, though even some of these are dying. Nor is reliance on bamboo a long-term solution: many bamboo species die en masse after a simultaneous flowering. At such

a time, storks that are over-reliant on bamboo will find no place to nest.

The dominant bamboo at Wat Phai Lom is reported to be *Bambusa arundinacea,* which is thought to flower at about 30 years of age. The Royal Botanic Gardens at Kew in England has records of this species flowering in 1963. Since all plants of the same species flower and die at about the same time, anywhere in the world, this may mean that Wat Phai Lom's bamboo clumps will die in 1993, precipitating a

In its search for ever-scarce building materials a stork (above) attempts to dismantle a section of the temple roof.

After hatching, young storks (right) spend 40 to 45 days before they fledge. When they leave the nest, other storks may move into it to raise another brood.

major disaster for the stork colony.

Plainly put, there now appear to be too many birds to survive at Wat Phai Lom. One theoretical solution—killing the excess—is unlikely to be tried. The practice would cause a public outcry in Thailand; monks would probably not agree to it within the temple grounds.

But the monks have shown that they are agreeable to other management schemes. Attempts have been made at tree-planting, but they have failed in the heavily polluted soil; the search for adaptable tree species continues. The birds were offered artificial nest platforms made of wire hoops, but they rejected these sites. In the late 1970s an attempt was made to start a new colony by capturing 37 young birds and releasing them in southern Thailand. Due to a

combination of stress and parasites, they died. One promising new idea: placing straw around the bases of the remaining trees. The straw could be periodically removed, together with its accumulation of droppings. Though only a palliative, and a labor-intensive one, it seems worth trying.

At present, the monks' tolerance of the storks is in the best Buddhist tradition: it cannot be very pleasant to live with the noise and smell, and the seasonal infestations of ticks that climb by the thousands up tree trunks in the early evening to feed on the storks. In fact, the ticks do not confine their attentions to storks; given the chance, they feed on humans too.

There is another temple whose grounds adjoin Wat Phai Lom and contain many suitable nesting trees. At present, though, there are no storks there. It is said that the other monks "don't like" the storks. There are ways of discouraging the birds, such as smoking them out, without having to kill them.

On my last visit to Wat Phai Lom I came across a young bird with a broken leg. I sought out the abbot, who collected it and took it to the "nursery," a small fenced area that already contained a bird with a broken wing.

"The young ones often fall down when there's a wind," he explained. "We put them in here and try to look after them. As good Buddhists we do what we can." He looked around for some snails that had also fallen to the ground, then gently opened the bird's bill and placed the pieces of food inside.

As I watched this act of kindness, I reflected that although it was touching evidence of the monk's very real concern for the welfare of a single bird, it did not begin to address the problems facing the colony as a whole. For this to happen, the modern sciences that call attention to the interconnectedness of life—such as field ecology and biology—must be joined to the teachings of The Enlightened One. ∎

Nigel Sitwell is a former editor of the British magazine Wildlife. *London-based photographer Michael Freeman visited Wat Phai Lom twice last winter for this story.*

Can Filipinos Learn to Love This Bird?

Western biologists trying to save the last 300 Philippine eagles learn to cross a huge culture gap

BY ROBERT S. KENNEDY

As part of a program to acquaint local people with the Philippine eagle, Ronald Krupa (below) displays a hooded captive to children in a remote mountain village. The endangered raptor (previous page) is one of the world's most powerful eagles.

Up in the treetops (right) author Robert Kennedy and photographer Neil Rettig (in protective helmet) visit a baby eagle in its nest 146 feet high. Feats like this were routine; any problems arose from cultural differences in a strange country.

Wolfgang Salb (2); pages 34–35: Neil Rettig

"A small boy was once captured by a monkey-eating eagle. The bird carried him to its nest where it kept him as its personal boy for five years. It brought food to him but would not let him go. When the boy was grown, he managed to hide under a waterfall and escape by swimming downstream under water. I knew the man—but he is already dead now."

—Alonso, leader of the Casiguran Dumagats, a vanishing tribe of Negritos in the Philippines

THE monkey-eating eagle is a huge forest hunter found solely in the Philippines on the mountainous islands of Luzon, Samar, Leyte and Mindanao. Today, this magnificent bird—now called the Philippine eagle—is so rare that probably fewer than 300 remain. I first came under its spell 11 years ago. Then, as a 24-year-old Peace Corps worker, I traveled to the remote mountains of the Philippines to help protect the species. But as Alonso's story points out, the local people's perception of the bird was far different from my own, and over the next decade, my experiences—shared by my wife Anne and various coworkers—pointed up the almost overwhelming difficulties faced by foreign wildlife biologists in many developing countries as they come to terms with huge cultural gaps. Studying eagles was easy. The hard part was coping with misunderstandings between ourselves and the people in the eagle's range.

Discovered only 87 years ago, the Philippine eagle has a crest of long, lanceolated feathers, piercing blue gray eyes, and a huge, highly arched but unusually narrow bill. Its status and habits were unknown until the early 1960s when a Philippine scientist, Dioscoro S. Rabor, alerted the world to its precarious status and pointed out three reasons for its decline: illegal logging and agricultural practices, which were destroying its forest habitat; trophy hunters who had the birds stuffed for exhibit in their homes; and zoos which captured the eagle for private display.

Although Rabor is now recognized as the father of conservation in his country, his early attempts to gain support from his compatriots fell on deaf ears. "We Filipinos have in the past not listened to one another when it comes to important matters like conservation. Rather, we will accept the word of inexperienced foreigners before we listen to ourselves," he explained.

One outsider who heard Rabor's pleas was Charles A. Lindbergh, the famed aviator. He visited the Philippines several times between 1969 and 1972 to help establish a conservation program for the eagle. A few months after his last visit, I arrived to assist the government in saving the species.

Although I was trained to study birds of prey, I had never been in a developing nation before, and I was soon dealing with indigenous mountain people quite different from myself. To them, the eagle was just another bird, probably less important than a domestic chicken. Nine months later, before I could devise a strategy to deal with this attitude, a religious war broke out, and I went home. But the stage was set for a subsequent visit. During that first trip I had observed how Filipinos love movies and, in particular, color photographs. As I left the country, I vowed I would return to develop an effective educational program using pictures.

Fortunately, in 1976 I met Neil L. Rettig, Wolfgang A. Salb and Alan R. Degen, three young naturalist-photographers who had just completed an extraordinary film on the South American harpy eagle. Together, we formed a non-profit organization, Films and Research for an Endangered Environment, Ltd., to fund our work, and in October 1977, we departed for the Philippines to study the eagle. Ronald E. Krupa, a specialist in bird handling, joined us to establish a breeding program for captive eagles.

Our first step was to learn more about the bird's natural history. Scientists did not even know when they bred or how frequently, where they nested or how to distinguish a young eagle from an adult. In six weeks we found our first nest, and by the project's end, 18 months later, we had discovered a total of five active nests and two inactive ones. That enabled us to answer the questions, and in the process we shot more than 50,000 feet of movie film.

At all nest sites, the eagles laid a single egg. During the incubation period, which lasts about 60 days, and the first third of the nestling stage, the males do

What looks like heavy artillery (below) is simply a 1000mm lens on a camera in a blind shared by author Kennedy, at the right, and photographers Wolfgang Salb, left, and Alan Degen. This fledgling on a snag (right) is six months old.

most of the hunting, while females remain within the nest area performing roughly three-fourths of the incubation duties and nearly all of the feeding of the young. Later, both males and females hunt and both take turns feeding their eaglets. The young we watched remained on their nest for up to 22 weeks after hatching; and, if successful, the eagles' nesting cycle lasts two full years.

Despite its early name, we discovered, the monkey-eating eagle rarely preys on monkeys. In descending order of importance, the eagles kill flying lemurs—nocturnal vegetarians that glide from tree to tree on flaps of skin as flying squirrels do; palm civets—nocturnal catlike mammals; flying squirrels; snakes—including the deadly Philippine cobra; bats—one weighed less than an ounce; rufous hornbills; monkeys and hawks. On one occasion a bird fed on the leg of a 30-pound deer.

The eagles are forest species and their prey are either forest or forest edge animals. We were therefore surprised to find that some nest sites were located where there was almost no forest of any kind. This gave us hope that the eagles could adapt to continued human pressures provided enough support forest was left for them and they were not shot or captured.

Additional hope for the eagle's future came when we discovered the birds on Luzon, Samar and Leyte islands where they were believed extinct or nearly so. The additional data helped us to roughly estimate the total population in 1979 to be between 300 and 500.

Our problems with people began when we arrived in the country, and they were especially evident with an indigenous local tribe called Manobos. Despite our efforts to keep these people involved in our activities, they kept asking, "Why are the Americans really here?" Seven months after we arrived, we discovered they believed we were looking for gold.

On the island of Luzon, the story was similar. While counting eagles in 1978, Anne and I spoke to mayors, loggers, barrio leaders and chieftains, always explaining how rare the birds are and encouraging people not to harm them. We believe our message was basically understood, but as it filtered out to the people, it quickly changed.

At one village, Casiguran, some locals believed we were searching for a huge eagle that had escaped from a zoo and that we wanted a reward for a valuable gold ring on its leg. Others thought we were members of the new People's Army, a group actively trying to over-

Feeding her single young, a mother eagle (below) belies her species' reputation for eating monkeys. Favored foods include flying lemurs, flying squirrels, snakes and palm civets. Until 1978, the bird was officially called the monkey-eating eagle.

This free-flying eagle (right) helped researchers determine the species' range and habitat. It was captured, tagged with a radio transmitter and released. Movies of birds like it are used in an education program to help save the eagle.

throw the government, and that our search for the eagle was a front. Still others reacted less to the bird than to me because they thought I was sickly. Two days earlier in a nearby barrio I had thrown up after eating lunch.

On a subsequent trip to Luzon, I almost abandoned our program because of other misunderstandings that led to the killing of two birds. Loggers, hearing that Robert Kennedy was in the area looking for eagles and believing that I was the late senator's son, captured one of the birds as a gift for me. They beat it to death soon after because they were terrified by its great strength. The second eagle was killed by a former headhunter of the Illongot Tribe. He had overheard me talking to his chieftain and was curious to find out what was so special about the eagle.

These were disturbing events to me, but I later learned they had educational value. After we talked with the men, they realized their mistake, so they agreed to spread the conservation word to their people. Between 1970 and 1978, five birds had been captured in Luzon and 11 shot. Since the two additional deaths, there have been no new reports of eagles killed in the two areas.

Our attempts to set up eagle sanctuaries led to other problems, however. At Kiandang on Mindanao, where we'd found our second eagle nest, the nest tree was scheduled to be cut by loggers. Fortunately the logging company agreed to pull out its equipment and to set the area aside as a sanctuary.

Eventually this area was recognized by the government. However, government officials made two mistakes that placed us and the eagles in great danger. By law, eagle trees are protected from human disturbance within a kilometer radius of the nest, but in this case, while we were away on vacation, the government decided to expand that no-man's-land to two kilometers. It informed more than 200 families living within the sanctuary that they would be relocated, and these people literally put a price on our heads. At one point they approached an American Catholic priest and possibly would have killed

him if he had not identified himself. "Are you with the eagle?" they had asked him.

After we negotiated on their behalf, the government finally agreed to restrict the sanctuary to a kilometer radius around the nest. Although those families living within that radius could retain their farms, no more land clearing would be allowed.

Neil Rettig and I, along with a government representative, brought the revised terms to Kiandang. As we approached the barrio meeting place, we were greeted by faces filled with anger and hatred. But as we explained they would not have to be relocated, their anger turned to happiness and they gave us a standing ovation.

The message we received from the people that day was clear. They view the remaining forest in the mountains as potential farmland. Their shifting agriculture was the way of their forefathers. To turn a large tract of forest in the mountains into an eagle sanctuary was like telling them that the eagle is more important than they are.

After that, we decided to shift our approach. Rather than try to preserve the forest for the eagles, we would try to preserve it for the use and well-being of people. With that in mind, we set out to produce two educational films. The first, carrying the simple message, "Don't harm the eagles or disturb their nesting places," would portray the growth of a young eagle and would stress the human characteristics of the species' family life. The second would illustrate the importance of the forest as a renewable resource.

The first movie, which was completed in 1981, became a great success. Tens of thousands of people have seen it on television, in schools and at public meeting places. The second film is still unfinished because of lack of funding.

For the eagle, problems remain, to be sure, and for the nearly four years that have elapsed since the completion of the filming, we have enlisted a number of Filipinos to help. We are training these people to succeed. To avoid misunderstandings such as the ones we en-

Wolfgang Salb (2)

countered, we let our new counterparts approach indigenous people alone.

Eddie Cañada, whom we had hired as a field technician to help us track eagles with radio telemetry, was one barometer to gauge how well we were doing. When we met him, he could not read or write, nor speak English. In the fall of 1981, Eddie and I and several members of our research team were sitting around a campfire in a remote mountain area of Mindanao. As usual, we talked about the eagle and how to save it. Suddenly, Eddie blurted out, "It's like we are fighting a war." To me his words symbolized a transition. Now his job was a personal cause, not just work. He had finally grasped why we were doing what we were doing. It was the start of a bridge across the cultural gap. ∎

This April Robert S. Kennedy joined the National Wildlife Federation as a senior biologist. He heads up the Raptor Information Center, which disseminates information about birds of prey. His eagle work was supported by the Marcos Foundation and World Wildlife Fund among other groups.

Shore Leave

After solitary years at sea, albatrosses come ashore in shifts, to mate and raise their chicks

*F*rans Lanting is a persistent fellow. The California photographer spent six months on America's Midway Island, getting close to many of the 500,000 albatrosses that nest there. The birds share their remote Pacific base with 200 U.S. Navy workers. Lanting's photographs, showing how the birds court, nest, fledge chicks and generally raise cain, may be the most revealing albatross pictures ever taken.

Eagle Imitation: *Casting an ominous shadow, a young Laysan albatross exercises its wings in the afternoon sun (above). As an adult, its wings will measure six feet tip to tip.*

Pilot Error: *More accustomed to life at sea, this bird was apparently attempting to land when it ran into a Midway Island tree (right).*

Junk Food: *Albatrosses sometimes get confused about what's edible. They feed their chicks small plastic objects like these (far right), arranged by the photographer in front of an adult.*

Photographs by Frans Lanting

The Great Debate: *Getting to the heart of the matter, this adult pair is performing a noisy courtship display known as "bill clappering."*

The Honeymooners:
Courtship among albatrosses often includes close contact (left) and mutual preening.

Late Bloomer: *Albatross chicks mature slowly (right); none are able to take wing until their sixth month.*

A Very Close Trim: *Nesting on earthen mounds (below), Midway's birds show little fear of their human neighbors.*

"I focus on Kenya's birds"

STORY AND PHOTOGRAPHS BY M. PHILIP KAHL

WE HEARD THE ROAR behind us, like a waterfall approaching at the speed of an express train. "Run!" my companion shouted. "Here comes the rain." It was mid-afternoon and the day had been clear and pleasantly mild. Biologist John Hopcraft and I were hiking back to my Land Rover after photographing storks, herons, ibis, spoonbills and other water birds of the Orongo herony along the shore of Lake Victoria, near Kisumu in western Kenya. When I turned, I saw a gray wall of cloud and water heading our way across the marsh from the southeast.

We pulled our equipment close to us and broke into a clumsy lope across the muddy ground, but after only a few steps, the downpour struck us full force. John grimaced with pain as the driving rain stung his bare legs between his shorts and his boot tops. When we reached the car, we were soaked and so were our cameras.

Torrential rainstorms and other minor discomforts notwithstanding, my family and I had fallen in love with Africa from the moment we had arrived. I had originally come on a two-year National Science Foundation postdoctoral fellowship to study the ecology and comparative behavior of the six species of storks that live there. For most of our stay, which eventually stretched to four years, we lived a mile above sea level and within one degree of the equator, where perpetual summertime prevails. In these areas, daytime temperatures rarely go above the low eighties and at night it's always cool enough for a fire in the fireplace and a heavy blanket on the bed. The seasons hinge mostly on the wet-dry cycle which, at the equator, repeats twice yearly, so that it is rarely extremely wet or dry.

Tropical regions like this one are fertile study areas for the field biologist because of the great diversity of animals living there — more than 1,000 kinds of birds, for instance, in Kenya alone. Such a variety exists probably because the creatures of tropical regions have had a long time to evolve undisturbed by drastic climatic shifts. That stability, in turn, enabled them to adapt to all available forms of food and cover.

At Kisumu, the breeding season begins in March or April, when the first heavy rains flood the shallow marshes adjacent to Lake Victoria, providing an abundance of accessible fish. I had built a tower to observe the three species of storks that nest specifically at Orongo, but that vantage point was also an excellent place to photograph the other birds in surrounding trees.

For wildlife photography, the conditions at Orongo are nearly perfect at that time of year — 12 hours of brilliant tropical sunlight and a dizzying array of spectacular subjects. As the weeks went by, I found myself behind my cameras more and more, and some of the results are reproduced on these pages. In all, I have taken more than 10,000 pictures of the birds of Kenya, but not all of those came easily. In fact, that sudden cloudburst was just another everyday occurrence in my quest to photograph Kenya's magnificent water birds.

During the previous dry season, I had made several trips to the Kisumu bird colony to construct the tower, which rose 30 feet near the nest trees. One night, midway through the construction — while I was camped in a small tent in the dry marsh — a crackling sound woke me up. At first, I thought it was an approaching grass fire, and I sat upright in my sleeping bag to scan the horizon. But the sky was dark. Then I realized that the crackling was coming from all sides and even from above and below.

Quickly, I beamed my flashlight through the mosquito-net window and saw that the tent walls were covered with "safari ants." There were thousands of these three-quarter-inch-long, reddish-brown creatures, and it was the sound of their snapping mandibles that had awakened me. Safari ants inflict extremely painful bites and, in an effort to discourage them, I slapped the insides of the tent and pounded on the floor. To my great relief, they all vanished within minutes — apparently having decided that there was nothing available here to eat — and I was left to sleep in peace.

Courtship behavior of African spoonbills is characterized by fighting between pairs (the two lower birds are females). Author Kahl took this picture not far from his observation tower at Kisumu, where hundreds of water birds breed after the spring rains start.

Where everything comes up pink

Often numbering more than a million, lesser flamingos (above) mass along the shore of Lake Nakuru where they consume an estimated 180 tons of blue-green algae daily. Although Kahl (left) photographed this flock from his Land Rover on shore, the picture of the greater flamingo family at Lake Elmenteita (right) was more difficult to get. As the parent dipped low to feed its offspring a regurgitated liquid, Kahl was shooting from inside a floating blind made from hollow drums covered with burlap. The naturalist, who spent four years in Kenya during one visit, has taken more than 10,000 still photos of wildlife there and countless feet of movie film.

To dry its plumage, a white-necked cormorant (above) spreads its wings. At Nakuru, the bird feeds on *Tilapia,* an introduced fish.

Foraging for frogs, the hamerkop (left) waits where a small stream crosses a road. A duck-sized bird found only in Africa, it is the only member of its family. Its huge nest may reach six feet high.

Alone in a tree, a sacred ibis (right) prepares to join a nearby courting group at Kisumu, where 50 to 75 pairs breed. The bird uses its bill to probe the mud for such food as shrimp and crayfish.

Within a few days, I had completed the tower with the help of several African assistants. At the top, we added a corrugated-tin roof for protection from sun and rain, a railing, and a pulley system to haul equipment to the "penthouse," where I would spend many days observing the birds that would soon be nesting at eye-level in the surrounding trees. Within 20 feet of my platform were nests of yellow-billed storks, African open-billed storks, black-headed herons, cattle egrets, African spoonbills, sacred ibis, great egrets and long-tailed cormorants.

Once the rainy season began, rain poured down almost every afternoon and the previously dry marsh — hard and drivable before — became a quagmire of sticky, black mud. During one of my early visits to the area, I thought it would be possible to drive to the tower, thus saving a long walk through the water with all my equipment. With more optimism than good sense, I splattered ahead in low gear and was soon hopelessly mired, axle-deep. That foolhardy move cost me the remainder of the day.

With five sturdy Africans digging ceaselessly for several hours, we slowly cleared a great, shallow bowl in the mud around the car. Then, hooking a team of oxen to a heavy rope, we tried pulling it free. The oxen strained, the vehicle did not move, and the rope broke. Again we set to digging. By late afternoon, we had moved an impressive amount of mud

and, just as the sun was setting, with six adult men and about 15 schoolboys hauling on the mended rope, we finally managed to tow the car to higher ground. On future trips, I carried my equipment to the tower.

Portages into the tower for a three- to four-day stay were a major operation. I generally hired several porters to help carry tent, sleeping bag, food, water and camera equipment. It was about two miles of hard going through soft mud and knee-deep water. Since the sluggish waters bordering Lake Victoria are heavily infested with the Bilharzia parasite, which burrows through the skin to infect the liver and bile duct, I wore hipboots to keep dry. When crossing areas too deep for my boots, I asked one of the heftier porters, who had already been exposed to the parasite, to carry me also. Although minor pieces of equipment were occasionally dropped in the mud or water, luckily I never got dunked myself.

During the time I was photographing the water birds at Kisumu, I was actually living in the town of Nakuru, about 100 miles to the east. Nakuru is located in the Great Rift Valley, 6,000 feet above sea level and about 30 miles south of the equator. Kenya's Rift Valley is studded with lakes, which are the home of some of the world's greatest water bird spectacles. At Lake Nakuru — only a few miles from the town in which I lived — there are frequently as many as a million flamingos which form a solid

pink fringe of birds around the shore.

On my first visit there, I arrived late in the day and made my way, after sunset, to a spot near the southern shore. I camped in my Land Rover on that dark moonless night and was kept awake by the constant roar of avian voices. In the darkness, I could barely make out ghostly shapes moving along the nearby shore. Occasionally, small groups would fly out or return to the flock, and I could hear them gabbling overhead.

In the first misty light of a cool dawn, I was up and straining for a glimpse of the lake. The light slowly increased, but the mist rising from the water seemed to thicken. As visibility gradually increased, I could see a solid mass of pink birds along the shore in front of me. They were lesser flamingos and so numerous that I couldn't begin to count them. The tightly packed flock stretched away into the lake and along the shore in both directions, until it disappeared into the mist.

As the rising sun burned off the mist, more and more birds were revealed. By 8 a.m., the day was clear and the entire lake was visible. As far as I could see, the water was covered with pink birds. Along the shore, they were packed tightly into a jostling mass several hundred feet wide. Farther out, in deeper water, birds were swimming and feeding in scattered groups. Here and there, among the shorter and redder lesser flamingos, were a few taller, paler birds, the

Its bill open, possibly to warn its mate of its arrival, a black-headed heron (above) flies back to its colony at Kisumu. Kahl considers this backlighted photograph one of his best.

Beak to beak, two courting yellow-billed storks (right) jockey for a potential nesting site close to Kahl's observation tower at Kisumu. The white deposits on their legs are caused by urine which the birds purposely direct there to cool their bodies through evaporation.

greater flamingos, outnumbered 100-to-one by their smaller cousins.

Although it is easy enough to view the masses of flamingos and other birds from the shore at Lake Nakuru, I needed to get closer for photos. Initially, I tried a slow approach in a small inflatable canoe, but the birds would not permit that sort of advance. On one occasion, I nearly drifted onto a submerged herd of hippos. When I was a few yards away, one suddenly raised its head and gazed at me with angry, pig-like eyes. I reversed the canoe in record speed.

Since the canoe was not producing the results I wanted, I was delighted when Leslie Brown, East Africa's resident flamingo expert, suggested a "floating blind," which rested on hollow drums and was covered with burlap walls and roof. I copied that design and added some refinements of my own. When the blind was ready, I put my cameras inside, crawled in and — wading on the bottom of the shallow lake and pushing the blind with me — headed for a cluster of flamingos. As I moved forward through the shallow, turbid water, I could see myriads of tiny organisms — which greater flamingos feed on — swimming up to the surface and then diving down out of sight again. They were so abundant that I could actually hear them making the water "fizz" like an uncorked bottle of champagne. At first, the birds at Lake Nakuru were wary of my floating blind, but later they seemed to be-

come more accustomed to it, and I was able to approach within range of my telephoto lens.

In addition to flamingos, there are many other water birds at Nakuru and other nearby lakes. Yellow-billed storks, herons and pink-backed pelicans fish in the shallows; great white pelicans and cormorants forage in deeper water. In the marshes where freshwater springs enter the lake, a variety of other water birds live. All told, more than 400 bird species have been recorded at Lake Nakuru, and I was able to photograph many of them from fixed blinds along the shore.

Even while living for several years in a comfortable apartment in Nakuru town, only a ten-minute drive from the lake, I never tired of camping near the shore at night. A favorite campsite was along the cliffs near the southwest corner of the lake. Some of my most memorable nights in Africa were spent there, listening to leopards coughing in the nearby forest, while flocks of flamingos called overhead as they circled the lake on their early evening flights. As the nightly chill increased, I would move closer to my campfire and think of the new richness of birdlife that the morning would bring. ∎

Ornithologist M. Philip Kahl, an authority on storks and flamingos, has visited Kenya twice since his original trip. This summer, he plans to go back again for more observations — and still more pictures.

56

Splendor at the Shore

PHOTOGRAPHS BY
CHARLES G. SUMMERS, JR.

ONLY one animal — a tiny brine shrimp — survives within its saline waters. But for birds, Utah's Great Salt Lake is truly a desert oasis. White pelicans nest in profusion on one of its remote islands, while along its eastern shore — where three rivers flow into the lake — a series of man-made dikes that was originally built to control flooding has created thousands of acres of shallow ponds and marshes. To these wetlands, some two million shorebirds, waders and waterfowl migrate each year. Like the striking creatures shown on these pages, most nest securely in the area's nine state and federal reserves.

Strutting in unison, two black-necked stilts parade along the lake's eastern shore, where a series of man-made dikes has created thousands of acres of wetlands. Such courting behavior is common for a pair of the gregarious, 15-inch creatures, which will eventually share all nesting duties after mating.

60

The Great Salt Lake's eastern shore is rimmed by marshes that offer abundant food for waders and shorebirds

Poking its head up from a nest of dead reeds, a hungry chick cries out as a male Forster's tern passes its catch to its brooding mate. In the fall, the adults' head feathers will turn white, except for small black eye patches.

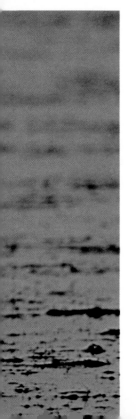

A famed "tattler," the greater yellowlegs is known for its loud, whistlelike calls. Wary and elusive, it will sound the alarm at even the slightest threat to its territory.

Tall reeds growing away from the shore form a protected nesting site for this snowy egret and its two fledglings. Gathering in colonies, the egrets often lace the reeds together like woven mats.

Though not saltwater birds, white pelicans nest in large numbers on Gunnison Island in the Great Salt Lake. From there, they must travel to nearby freshwater lakes, where they fish in groups on the surface. After gorging themselves, they return to the remote island to feed their young.

Avocets build their nests in open areas along the shore, enabling both parents to watch out for skunks and other predators. Within hours after the speckled eggs hatch, their precocious chicks will be capable of swimming and feeding themselves.

COOL IN CALIFORNIA

Penguins have found a congenial new home in San Diego, and scientists have found a living laboratory

By Vic Cox

JERRY ROBERTS (SEA WORLD)

BIOLOGISTS have to go where the action is, and to study some species of penguins, that means going to Antarctica," says Frank Todd. The 41-year-old, red-headed chief curator at Sea World in San Diego, California, should know. Since 1972, he has made a dozen trips to the southern tip of South America and Antarctica—frequently braving 100-mile-an-hour winds and below-freezing temperatures—to learn more about the flightless birds.

Today, as a result of those efforts, other scientists can now study penguins in relative comfort in Southern California. The site: Sea World's new Penguin Encounter, a unique, multi-million-dollar structure that is the largest penguin breeding and research facility in the world.

"We are trying to create more than exhibits here," notes Todd. "We are striving to establish ecosystems." Those ecosystems include outdoor accommodations for Humboldt penguins from the temperate waters off Chile and Peru and refrigerated indoor habitat for several other species of penguins and northern cold-climate sea birds called alcids.

For eight years, Todd and his associates housed populations of the birds behind the scenes at Sea World, hoping to determine whether or not the creatures could be bred outside their native habitat. The researchers' studies eventually paid off. "We've now achieved so many records that the concept of 'impossible' no longer exists for us," says Todd. Those records include establishing the first self-perpetuating populations of emperor, king, Adelie and endangered Humboldt penguins in captivity, and producing the first emperor and king penguin chicks ever raised by hand to fledgling state. Last spring, some 300 of the birds took up permanent residence in Sea World's new cold-climate exhibit.

For visitors, entering Penguin Encounter is a chilling experience. A blast of cold air hits them as they pass giant photomurals of penguins in Antarctica, massive glaciers, icebergs and ice floes lined up like some great white fleet from the past. It is Todd's way of shaking arid Southern California from the minds of the birds' visitors. "I want them to feel something of the penguins' environment," he says.

The environment is meant to be

equally convincing to the birds. Some 10,000 pounds of finely crushed ice is blown onto the exhibit's rocks and ledges each day; a whale's jawbone is set into a 100-foot-long ice shelf. The air, which is filtered to remove germs that the Antarctic birds have no immunity against, is kept well below freezing, and the seven-and-a-half-foot-deep pool of seawater is kept at a chilling 45 degrees. Adelie, emperor, king, gentoo, macaroni and rockhopper penguins all share this pseudo-Antarctica in apparent toleration.

To assist researchers, three remote-controlled television cameras cover the entire length of the exhibit, thus allowing scientists to zoom

Curator Frank Todd (above) has brought together six species of penguins in a new, sub-freezing habitat at Sea World in California. Hatching beneath its parent, this emperor penguin (left) is one of the first such chicks born outside of Antarctica.

in and videotape any behavior they wish. Another interior display offers views of alcids as they roost on artificial cliffs and rocks or dive for food. Anchovies are periodically released into the 19,000-gallon pool, so visitors can see the seabirds zip under water in live hunts. Eleven species of alcids live here, including the horned, tufted and Atlantic puffins, common murres, rhinoceros and Cassin's auklets, and pigeon guillemots. Unlike the penguins, the alcids can fly, so their skylit enclosure gives the illusion of space.

The penguins are kept on Antarctica's dark and light cycles. Since the Southern Hemisphere's seasons are re-

versed in San Diego, that means that emperor penguins, which breed in Antarctica's winter darkness, are laying eggs and brooding chicks under low light conditions in the Encounter exhibit at a time that coincides with the height of Sea World's summer season.

Some scientists were concerned that increasing the light so that visitors could see the penguins would interfere with mating, but the Encounter's first emperor breeding season has allayed the anxiety. Last year, seven emperor eggs were laid and two hatched, which is about the average for a group of relatively inexperienced penguin couples, according to assistant curator Frank Twohy. Both of the new emperor chicks had to be taken from their parents and hand-raised because the adults weren't feeding the infants properly.

In the wild, emperor and king penguins rotate their eggs every few hours during incubation. Working at Sea World, two zoologists from the University of California at Los Angeles, Terry Bucher and George Bartholomew, discovered that the birds do this because otherwise the eggs will not be evenly warmed. At UCLA, the researchers used special plastic eggs with implanted biotelemetry cores which measured various incubation temperatures. Todd and his staff then used the resulting information as a guide for setting the temperature on their

mechanical incubator at Sea World.

Bartholomew, one of the nation's leading authorities on sea birds and marine mammals, considers Sea World's penguin project an invaluable asset for researchers. Because he conducted his initial studies on eggs there, for instance, the biologist completed his subsequent 1982 Antarctic field work with less equipment and in half the time it otherwise would have taken. "You can't go to San Diego to find out how far penguins will go in search of krill, which is one of the things I studied in Antarctica," says Bartholomew. However, he adds, nesting behaviors, male-female relationships and most aspects of penguin physiology can be researched in San Diego, and probably with greater accuracy than in the wild. "You don't need to work in freezing rain to study embryos," the scientist points out.

Bucher stresses the complementary nature of information from field studies and from self-perpetuating groups of captive penguins. And Frank Todd concurs, noting that "much of the success we have achieved at Sea World can be attributed to information gathered in the field. Conversely, much of the information obtained from the captive colony has been applied to field programs."

For example, the age of various species of penguin chicks can be determined in the field from their weight because of age-weight relationships charted by Sea World aviculturists. Most research done at Sea World since 1972, when the National Science Foundation called Todd and initiated the project, has focused on captive breeding. (Todd says that Adelie penguin chicks in the project now experience less than a 10 percent mortality rate, compared to a maximum 70 percent death rate in the wild.) But some studies have been aimed at determining the sex of birds with similar appearance through hormone analysis, the meaning of emperor penguin calls and the energy involved in flying underwater.

All of which delights Frank Todd. "From the beginning, the whole idea behind the penguin project was research," he says. "It's a dedication we won't relax." ∎

Southern California writer Vic Cox met his first emperor at Sea World last summer.

Big Red booms in Cincinnati

By one measure, this Ohio city is the cardinal capital of the country

BY STEVE MASLOWSKI

ONE wintry afternoon a few years ago, I was walking with a friend in a neighborhood of my hometown, Cincinnati, Ohio. By chance, we came upon a small field, about the size of an average backyard, that was filled with ragweed bent to the ground by heavy snow. There, we witnessed an amazing spectacle: nearly 150 cardinals had

Decorating a tree in Cincinnati, a flock of cardinals brightens up the barren winter landscape. According to annual Christmas bird counts, among the nation's larger cities, the Ohio community boasts the greatest concentrations of the songbirds. The red male (above) offers a striking contrast to his paler female counterpart.

flocked to the field, and they were busily gathering seeds. To me, the scene was very festive, not only because it represented the greatest concentration of cardinals I had ever seen, but also because Christmas was only a couple of days away. The brilliant red males stood out vividly against the snow — as if someone had spilled several bushels of apples. Perhaps that's why my companion — an avid birder — described the setting as "delicious."

In the early 1800s, it would have been difficult to witness such a spectacle in Cincinnati. About a hundred years ago, cardinals were rarely seen in Ohio; indeed, they were a rare sight anywhere in the United States, except in the South. But by clearing forests and altering

other natural habitats, we humans opened up many new areas to the adaptable red birds. As a result, in recent decades, cardinals have increased their range and, apparently, their numbers. Today, these songbirds flourish in many northern U.S. cities — though none of those communities can match our claim: according to the annual Christmas bird counts, among the nation's larger cities, Cincinnati is by far the cardinal capital of the country.

At one time, Cincinnati was located at the northern limit of the cardinal's range. The 1886 American Ornithologists Union (AOU) checklist shows that the species was only a casual visitor north of the Ohio River Valley. But shortly before the turn of the century,

the cardinal began a dramatic expansion. By 1910, the AOU checklist showed the species extending into southern Ontario in Canada, and parts of the Hudson River Valley. Eventually, the cardinal also spread west into Texas and South Dakota, and now, it even breeds in Manitoba and Nova Scotia.

Thus, Cincinnati currently sits more or less in the middle of the cardinal's U.S. distribution. Our city, in fact, is located near seven states that have chosen the creature as their official bird: Illinois, Indiana, Kentucky, Ohio, Virginia, West Virginia and North Carolina. No other species of bird has been similarly honored so many times.

The cardinal has three important characteristics that help to make it one of the country's best loved birds. First, it is distinctive. The brilliantly plumed male is relatively large (7½ to 9 inches) for a songbird. He is also the only all-red crested bird in North America, and thus, he is readily recognized even by small children. The female, about the same size as the male and also crested, is more subdued in color, but she also sports reddish hues. Second, cardinals sing well. Researchers have recorded at least 28 songs and calls for the species, and both males and females sing throughout the year. (The female's song, however, is much softer and less varied than the male's.) Third, the birds thrive around people. And because many cardinals stay in one place year-round, their entire life cycle can be easily observed.

In Cincinnati, during the first warm, sunny days of March, the cardinals begin their courtship ritual. The male feeds his mate, first cracking the shell of a seed, then offering her the meat. This goes on throughout the period that the female is busy building her nest — an open, deeply cupped structure lined with soft material.

Usually, the female places the nest only a few feet off the ground in dense shrubbery. And as long as such shrubbery is available, she is not especially concerned with how close to a house the nest is located. My father Karl, who has photographed wildlife professional-

Honored by seven states as their official bird, the brilliant male cardinal is the only all-red crested songbird in North America. In recent decades, the highly adaptable creature has steadily expanded its range northward into some areas where it was rarely seen a century ago.

Hardly birds of a feather, these newborn cardinals and American robins hatched side by side in the same Ohio nest (left). The four adults shared parental duties, though each species was more inclined to take care of its own kind. While a female cardinal broods the young (below), a male robin brings food to the nest.

ly for most of his adult life, likes to recall the first cardinal nest he ever saw — some 45 years ago. The female had placed the nest in a patch of shrubbery that was planted to hide a bathroom window. As a result, my father had a wonderful view directly into the nest, as long as he stood on the toilet.

A typical cardinal clutch contains three or four bluish white, brown spotted eggs, which hatch after the female has incubated them for about 12 days. The yellow brown fledglings are fed insects by both parents. The instinct to feed their young is apparently very strong — so strong that the adults may occasionally offer food to any young bird in the area.

I once found a cardinal nest that was being used simultaneously by a pair of adult robins. It was quite an experiment in integrated housing, because cardinals belong to the finch family and robins are thrushes. Nevertheless, three young cardinals and four newborn robins were all reared together. The four adults shared parental duties, though each species was apparently more inclined to take care of its own brood.

Normally, about ten days after they hatch, young cardinals are ready to make their first flight. Shortly thereafter, the mother leaves the fledglings and begins building a new nest. This will house the second of the two or three clutches of eggs that she will lay over the summer. The male bird usually stays with the young birds longer, however, and teaches them to find food.

Only about one fledgling in three lives long enough — about a year — to achieve its full adult plumage. Predation and accidents take a heavy toll that first year. But, as with most songbirds, mortality is high even among adult cardinals. An estimated 40 percent of the adult birds die from a variety of causes each year.

Despite this high turnover, cardinals are prospering in many parts of the country. This is due in part to the popularity of backyard feeders that homeowners stock with sunflower seeds. It is not unusual to see a cardinal eat a dozen sunflower seeds in one sitting, and Art Wiseman, an enthusiastic Cincinnati birder, saw one bird consume 27. Wiseman believes that the seeds may be a staple for some of the birds. He has kept an injured cardinal for eight years (with proper permits) and says, "Once in a while the bird will lay off some of its favorite foods, such as pyracantha berries, multiflora hips and corn, but it will go for sunflower seeds day after day." These days, residents of Cincinnati alone buy some 400 tons of sunflower seeds per year, and this figure is probably duplicated elsewhere.

In addition to helping the species by establishing feeding stations, we humans have also helped the birds by altering some habitat — often to the detriment of other creatures. At one time, much of the northern United States was covered by forests, which were good habitat for wild turkeys and pileated woodpeckers, but not for cardinals. By

---

The instinct to feed their young is so strong that the adults may offer food to any young bird in the area

---

the early 1900s, however, many of those forests had been felled and, in some cases, replaced by meadows and brush — ideal habitat for cardinals.

Because cardinals are nonmigratory, they apparently expand their range only when population pressure forces individual birds to explore nearby unpopulated territory. Habitat and feeding stations have both played important roles in this exploration. According to Jay Sheppard, an ornithologist with the U.S. Fish and Wildlife Service in Washington, D.C., nearly all the birds that pioneered in the north and west "probably explored the river bottom first, where they found the thicket type of habitat they like. When they encountered a feeding station near one of those valleys, they stayed put. Then, in spring, they bred. Possibly a founding group became established around the station."

Cincinnati's claim of being "cardinal capital" rests on the Christmas Bird Count, which has been sponsored by the National Audubon Society every winter for the past 82 years. On a designated day within two weeks of Christmas, more than 30,000 bird-watchers all across the nation — about 150 of whom live in Cincinnati — participate voluntarily in this census. Each locality has a 15-mile radius within which the count takes place. The volunteer counters go to this area at midnight and move through it either on foot or by car, recording every bird of every species they see or hear until the following midnight.

Since 1950, in every year but one, Cincinnati has reported more cardinals in the Christmas Bird Count than any other locality. (It was surpassed by Seneca, Maryland, in 1975.) Over the last 15 years, from 2,000 to 3,700 cardinals have been counted annually in Cincinnati. That bird count also provides evidence that the birds are well entrenched in the Ohio community.

One indication of this came after the winter of 1976-1977, the worst winter in the city's history, when the lowest temperatures and deepest snow levels were recorded. Several species of southern birds, which have moved north in recent decades, were seriously depleted by the hard weather. The Christmas Bird Count tally of the Carolina wren, for example, dropped from a national high of 600 in the previous year to 6. But Cincinnati continued to lead the country in overall numbers of cardinals.

Other bird counts indicate that some communities in the South may actually have higher concentrations of cardinals than Cincinnati does. The Breeding Bird Survey coordinated each summer by the Patuxent Wildlife Research Center in Laurel, Maryland, shows that the species is still very abundant in the South, and that a number of localities there have more cardinals per square mile than does Cincinnati. The leader is Hebbronville, Texas, southwest of Corpus Christi.

Whatever the numbers, the fact remains that the species has apparently developed a strong attachment to Cincinnati and its environs in recent years — an eloquent testimony to the cardinal's unique adaptability. ∎

Ohio writer Steve Maslowski coauthors a weekly nature column for the Cincinnati Enquirer.

71

This Raptor Keeps A Low Profile

And with good reason — the red-shouldered hawk is vulnerable to attack by more powerful birds

BY FRANCES HAMERSTROM

MENTION birds of prey — hawks, owls, eagles and falcons — and the first thing most people think of is wings. Conversely, the last thing most people think of is feet. But once you've experienced the exquisite agony of a raptor's talons piercing your flesh — as I have — you never again overlook the importance of those weapons. Raptors soar on their wings and tear with their beaks; they *kill* with their claws. Their feet, therefore, are critical to their survival.

In this respect, one bird of prey, the familiar red-shouldered hawk, has not been richly endowed. Its feet are small and weak. Consequently, the red-shoulder must be content with relatively small quarry. But that's not all. Because its puny feet offer an ineffective deterrent to attack by great-horned owls and other large raptors, this bird cannot swagger boldly through open country like its cousin, the red-tailed hawk. Instead, for its own safety, it must maintain a discreet profile deeper in the woods.

Anyone who is familiar with raptors knows about the red-shouldered hawk's weak feet and, as a lifelong ornithologist, I've had some firsthand experience with them. One day, I was holding one of these creatures on my bare hand, trying to get it to pose for a picture. Even untamed hawks ride a falconer's glove gently. But if a dog rushes up, or if a mouse squeaks, a large raptor may plunge its claws deep into the meat of the hand inside that glove. That doesn't happen with the red-shoulder. When my photographer friend imitated a mouse squeak to get our model to pose, the red-shoulder's talons gripped my hand — but drew just one tiny drop of blood.

Because of this, the red-shouldered hawk has found its niche in small ravines and river-bottom woods, where smaller prey is abundant and larger raptors seldom hunt. It often shares its habitat with the barred owl. These species have been known to nest within five feet of each other, and they hunt the same food — but at different times of the day.

Many birds of prey kill creatures larger than themselves. Red-shoulders do so only rarely. Once in a while, they latch onto cottontail rabbits and, after tumbling end-over-end, make successful kills. But usually, they pursue less spectacular game: mice, shrews, grasshoppers, spiders, small snakes and a few birds.

Red-shoulders nest near water, and frogs always seem to be part of their diet. Even so, the art of frogging does not always come easily to young hawks. One orphaned red-shoulder that I raised to maturity quickly found that frogs jumping on land were too unpredictable to catch, but that she could grab them during their straight underwater getaway. After each catch, she floated on the water with wings outspread — and each time I had to wade in to rescue her!

During courtship, the red-shoulders' low profile is forgotten for a short time and the birds soar high above a woodland, sometimes making power dives almost to the treetops. Their piercing calls, *kee-you kee-you,* advertise a likely nesting territory. When mating flights are over, the woods are quiet again, although some individuals call loudly when an intruder comes near.

The nesting birds shown on these pages are members of an insular race that resides in the southern third of Florida. There are four other races: three farther north in the East and the Midwest, and one all by itself out along the West Coast. All of these birds have darker, larger heads than the southerners.

As soon as the eggs hatch, the female shades her small young from strong sunlight, but she mustn't overdo it. Rapidly growing chicks need sunlight to ward off rickets, a disease that produces bent, distorted bones. Somehow, the female red-shoulder works out the delicate balance: sunlight, but not too much.

Red-shouldered hawks regularly add fresh green twigs to their nest (photo, page 22). They may do this for several reasons: to keep the young cool, to keep the top of the nest cup clean, and to protect the tender young hawklets from maggots and carrion beetles that feast on food remains in the lower layer of the cup. These beetles and maggots can attack newborn chicks through their thin belly skin and under the wings. After several weeks, however, the increasingly alert young hawks are snapping at flies and they find crawling beetles easy prey.

It's hard to say how red-shouldered hawks have fared over the years. For that matter, it's hard to say how hawk populations in general have fared. Although some species reportedly have been declining since the 1920s,

Blending into her surroundings, a female red-shouldered hawk shades her nest and chicks from the bright Florida sun in Everglades National Park. Smaller than the red-shoulders that range in the northern U.S., the Florida race begins nesting in early February, before the cypress trees are in leaf.

Wings outspread, the busy female prepares to carry a fresh green twig back to the roost. While brooding her young, she will regularly replenish the nest with branches from nearby trees. This activity, many biologists speculate, helps keep the crowded nesting area clean and the chicks cool.

other species are probably increasing.

Whenever a species seems to be declining, there is usually a fashionable scapegoat to blame. In the 1930s, and even more recently, many farmers systematically shot every hawk they saw "to protect the poultry." All across America, dead hawks were hung in rows along fences or nailed to barn doors. And egg collecting was still a popular hobby then. So eggers and shooters were scapegoats for apparently declining hawk populations.

Arthur Cleveland Bent, the famous ornithologist and author of *Life Histories of North American Birds,* was fond of posing a curious hypothesis. He believed that, in New England, many families of red-shouldered hawks were actually *saved* by the taking of eggs in April. After the first eggs were taken, he argued, the birds produced a second clutch in May. And those eggs stood a better chance of producing a brood when the leaves were out. Surrounded by the new spring foliage, the nests were harder to find and farmers were busy in the field and ceased looking for them. So the early eggers, according to Bent, not only saved some hawk families from shotguns, they also left behind carefully documented egg collections. These enabled modern researchers to associate eggshell thinning with DDT and the decline of the peregrine falcon, the osprey, the bald eagle and some other raptors. Bent, it should be emphasized, was a noted egg collector himself.

Distinguishing between handy scapegoats and real causes takes years. The killing of hawks invariably sparks strong emotional reactions from the public — but it is seldom the cause of a progressive population decline. A more serious but less recognizable culprit is destruction of habitat. A careful study in central Maryland showed that where habitat was left intact at the Patuxent Wildlife Research Center, there were actually more red-shoulder nests in 1971 than there had been in 1943! But where habitat had been disturbed by development, there were fewer.

Long-term hawk population trends, and the reasons behind them,

Defending her nest, the aggressive mother swoops down to intimidate an intruder. The photographer received the same reception each time he entered her territory. Though a powerful, long-distance flier like all buteos, she will rarely fly beyond the sight of her young chicks — even if her mate is in the area.

After tearing apart its flesh, the female passes a piece of freshly-caught bullfrog to a hungry chick. Because their small feet can't handle large prey, red-shouldered hawks hunt only in those areas where smaller creatures are abundant, and where larger raptors are less likely to range.

are difficult to chart because information is so hard to come by. It takes real ingenuity just to dig up reasonably reliable evidence of causes and effects, much less genuine scientific data. Not long ago, Stanley Temple, a professor of wildlife ecology at the University of Wisconsin, demonstrated the kind of sleuthing that is required. For many years, a series of bird seminars was held at Cornell University, a mecca of ornithological research in the U.S. At those meetings, members of the audience routinely reported on how many different kinds of birds they had seen since the last meeting. This struck me as merely a quaint exercise but records were kept of the proceedings and Temple, with his wife Barbara, had the wit to go through them.

From this research, the Temples identified ups and downs in numbers of hawks that roughly paralleled the quantities of DDT and other pesticides in use at various times.

But neither the cause of the decline, nor the subsequent increase, is known for sure. Hawks that eat food contaminated with DDT tend to lay thin-shelled eggs. But mysteriously, a thinning of only five percent showed up in the eggs of red-shoulders. My hunch is that many adults did not breed at all. To the best of my knowledge, no one kept track of individual red-shoulders during those critical years beginning in 1950.

Interest in what hawks eat is increasing today. Although the use of DDT is restricted now, some researchers fear red-shoulders may be picking up new pesticides.

Early food-habit studies tended to be conservation oriented. Researchers liked to prove that the hawk was "the farmer's best friend" by stressing how many mice and how few chickens they ate. In 1935, the red-shoulder was rated as "distinctly beneficial" by John May, author of *Hawks of North America.* The stomachs and crops of 444 red-shoulders that he dissected contained mammals and only 7 contained poultry.

For its stomach contents to be studied, a bird obviously must be dead. This has certain disadvantages from a research point of view. Consequently, modern researchers have developed more sophisticated techniques for finding out what a living hawk or owl has eaten. Every day a bird of prey swallows undigestibles. Digested residues are excreted — mostly as "whitewash" — and the undigestibles are coughed up as pellets, usually daily.

Undigested bones and owl pellets are easily identified. But hawks — including the red-shoulder — have strong digestive juices, and analyzing what they have eaten from their pellets, picked up in the nest, is the road to despair.

There is another way, known as "choking up" — a simple but delicate operation that untrained laymen should *not* try. You can tell if a hawk has eaten recently by a bulge under its chin. If the bulge is big, it is known as a "full crop." To find out what is in the crop takes only fingers — used very gently — to guide the contents out.

Every morsel that I can identify is slipped right back into the crop, but unidentifiable bones, fur or feathers are kept for study. It often takes many hours to determine what the bird actually ate.

Even if we knew all about a bird's diet and life cycle, there would still be an enormous gap of knowledge about its behavior.

Though my husband and I have studied birds in the field, and rehabilitated many injured birds of prey in our backyard, we soon learned that the more we discovered about their behavior, the more we had yet to learn. We were reminded of this one more time last summer, when we reared two red-shouldered hawks and watched them grow up just outside our Wisconsin farmhouse. By mid-July they were well grown, lively youngsters and any day we expected to see them making substantial kills on their own.

Just then, an unexpected present arrived: three orphaned, helpless little marsh hawks. I fed them and put them on the porch, hoping that the screen was strong enough that no red-shoulder could break through it and kill them. Not long after, I was alerted by a ruckus. The female red-shoulder was bumping at the screen and it took me a moment to figure out what she was up to. This bird, still a year or two away from sexual maturity, apparently was trying to feed those marsh hawk chicks a mouse! ∎

In addition to studying birds of prey, Fran Hamerstrom and her husband, Fred, have worked for decades to restore habitat for the prairie chicken in central Wisconsin.

The male red-shoulder returns to the nest several times each day, but usually stays there only long enough to pass his catch — in this case, a young black swamp snake — to his mate. At most other times, the male stays out of sight, hunting and protecting his favorite territory from other hawks.

Alert and anxious for its parent's return, a two-week-old hawk peers up from a nest lined with fresh wax myrtle and cypress branches. Its first flight attempts will begin in another three or four weeks. Red-shouldered hawks build their nests in tall trees, 30 to 60 feet above the ground.

Meet the real nesters of Australia

Almost all of Australia's land birds are full-time residents; they've had to adapt to the hot and often arid summers which discourage visiting migrants

STORY AND PHOTOGRAPHS BY MICHAEL MORCOMBE

THERE'S SOMETHING STRANGE about the land birds of Australia. They all live there! Except for two cuckoos and two swifts, the country has no migrant land bird visitors, and even these do not nest on the continent.

Compare Australia with other southern continents, which seasonally host millions of northern birds, many of which must fly thousands of miles to reach them.

Why have migrant land birds ignored Australia? The experts say the answer may lie in hot, often arid, and generally inhospitable summers. It follows that the full-time residents of the continent must have made excellent adaptations to their environment to thrive.

And thrive they do. The birdlife of Australia is famous for its uniqueness (many species are found nowhere else on earth), and for its splendid plumage. The best time to observe this parade of color is the mating season.

The first priority in the mating season is to stake out a territory. Regardless of the area's size, which may be an acre or more for some songbirds, the resident male declares it his own and defends it with the aid of his bright, boastful displays.

Once the male has staked out his claim, he looks for a mate, attracting and keeping her with the same brilliant displays he used to fend off rival males.

The first major activity of each pair is to set up housekeeping. Nest building work may be shared, or done solely by the female or the male, depending on the species. Many birds favor bushes, tree hollows or branches, while some dig their own nest tunnels in the ground.

Nest construction varies as much as location. Tree nests range from huge stick platforms to tiny woven bags to the silver-dollar-sized nest of the lemon-breasted flycatcher. Birds use whatever the environment offers, including grass, moss, leaves, branches, roots, seaweed, spiderweb, mud, fur, and even wire and human hair.

Birds have been observed stealing nesting material from other birds' nests. A few go so far as to scare out the builder of a finished nest and take over the new structure, although birds more commonly renovate abandoned nests.

When the nest building is complete, the female lays the eggs and begins incubating, which varies among Australian species from eleven days to two months. The busiest time of nest activity begins after hatching. Feeding, defending and sheltering the newly hatched requires most of the parents' energy. The survival rate of Australian broods is high because of the parents' defense against predators like the kookaburra and butcherbird, and because of nest camouflage.

Very often the species in which the males are most brightly garbed find the female left with most of the family chores . . . he has departed to find another mate.

But bright plumage may serve still another function. By attracting the attention of predators, gaudy males may save the better-camouflaged females and young.

When a brood leaves the nest, many parent birds prepare to raise another, but at the end of the season, which may last six months, all birds temporarily cease their family activities. Most birds show signs of wear, including dull-looking, bedraggled feathers. Now they moult and acquire fresh plumage for the next season.

The golden plumage of a rufous fantail brightens the gloom of leafy forests. Preferring the fern gullies and rain forests of Australia's northern and eastern coasts, the fantail houses its young in neat, wine-glass-shaped nests.

Nesting materials range from grass, leaves, mud, fur and spiderwebs to wire and human hair, and bitter competition may arise over tree hollows

The young of the scarlet robin, left, take advantage of their protective coloration by keeping perfectly still until they feel the touch of a parent's feet at nest rim. Australian robins camouflage their cup-shaped nests with bark, mosses and lichens, binding them with matted spider webs.

The rainbow bird may have evolved the near-perfect protection for its young, lower left. Instead of building a traditional nest, it drills a tunnel into the ground where it lays its eggs and raises its chicks.

A bird of the treetops, the striated pardalote, below, feeds on small insects and brings them to its domed nest of bark inside a small knothole. The camera's electronic flash froze the action of this little bird as it stretched out its legs to cushion the impact of landing.

Sweeping in for a landing, a crimson rosella is about to feed its young in a tree hollow, right. Wing feathers and tail spread to check the rosella's speed or propel it away from the tree. A member of the brilliant parrot family, the rosella inhabits coastal forests.

Australian robins exhibit strong pair bonding. During courtship this male red-capped robin brings gifts of food to his mate, and continues to supply her with food while she incubates the eggs

Nest-building work may be shared by a mated pair or be done exclusively by either the male or the female

The yellow-plumed honeyeater builds a fragile, open-cup nest of grasses, left, in the outer foliage of small trees. An aggressive creature, it will chase away other birds feeding in trees.

One of many species that survive in the sparsely vegetated, semi-arid interior of the Australian continent, the orange chat, lower left, conceals its nest in low bushes around inland salt lakes and marine marshes.

Ready for a takeoff, the male red-backed wren, below, mates with a less colorful brown female. He assumes a full share in feeding the young in a domed nest of dead grasses close to the ground. Preferring tall coarse grass and thick undergrowth, often in damp places, this wren inhabits tropical northern and northeastern Australia.

The western spinebill honeyeater, right, is not exclusively a vegetarian, and often captures large insects in the air. The pollinating honeyeaters have caused many plants, such as this red-and-green kangaroo-paw, to evolve flowers specifically designed for birds' transfer of pollen from flower to flower.

BIRDS OF A DIFFERENT BILL

There's nothing similar about the fortunes of these two woodpeckers

ALL TOO OFTEN, people confuse the two birds for one another and, in some respects, they are similar in appearance. The ivorybill and pileated are the largest woodpeckers in North America, averaging 20 and 18 inches respectively. Traditionally, both ranged widely through many eastern forests, and they ate similar foods. Why, then, did the ivorybill vanish, probably forever, while the pileated continues to thrive in areas the two creatures once shared?

The answer may lie in the subtle differences in their anatomy—and in their abilities to adapt to changing conditions. Neither bird, of course, reacted well to the logging of virgin forests in earlier times. The ivorybill became particularly vulnerable because it lived primarily in mature forests, amid trees at least 150 years old. In 1935, in Louisiana, ornithologist James T. Turner took the rare photograph shown below of one of the last nesting pairs of ivorybills seen in North America. In contrast, the pileated began reappearing in greater numbers in this century, apparently because it could survive in less mature woodlands and eat a wider variety of foods.

Though the ivorybill remains on the U.S. Endangered Species List, most biologists believe it is already extinct. "Tantalizing reports of the ivorybill's existence usually turn out to be pileateds—or hoaxes," says Jay Shepard of the Office of Endangered Species. "The pileated has now become so common that it is turning up at many backyard bird feeders." ■

Unable to adapt to changing conditions, the ivorybill (below) disappeared while the pileated woodpecker (right) flourished. This rare photograph of the ivorybills was taken a half-century ago on black-and-white film, and later hand-colored.

BLUEPRINT FOR
Bluebirds

By Eileen Lanzafama
Photographs by Michael L. Smith

A tenacious band of volunteers is engineering the songbird's return to its former range

URING a fit of spring cleaning two years ago, Cheryl Dodenhoff found an old wooden birdhouse buried in the back of a closet. The Long Valley, New Jersey, homeowner dusted it off and nailed it to a post in her backyard. Then she forgot about it—that is, until a bird watching friend visited her home a few weeks later. "My God!" her friend exclaimed, looking out a window. "You've got a bluebird!"

A generation ago, the presence of an eastern bluebird in Dodenhoff's backyard would hardly have produced such a response. As recently as the 1940s, the bird was commonly seen in the fields and yards of farms and suburban communities throughout much of the eastern half of the country. After World War II, however, the situation began to change. The eastern bluebird, plagued by a series of harsh winters, habitat loss and increasing competition with more aggressive birds for available nesting sites, began to disappear. "When I was a boy, bluebirds were almost as common as robins," recalls Lawrence Zeleny, 79, a retired biochemist from Maryland. "Now, most people under 30 have never seen one."

Experts disagree over the extent of the bluebird's decline. Some people believe their numbers have been reduced by as much as 90 percent during the

Bluebird house construction is booming these days. In northern New Jersey, retired school teacher Junius Birchard (above) has distributed some 6,000 nesting box kits during the past six years. Many bluebirds do not migrate and during a cold snap, adults may crowd inside a box for warmth (right).

past half-century; others think that some of the birds have merely become less visible, shunning heavily populated areas in favor of more remote habitats. No one knows for sure. However, one thing is certain: the little bird has been sorely missed. In recent years, its absence from traditional nesting areas has resulted in an almost unprecedented voluntary effort by hundreds of concerned citizens.

Some people, like Cheryl Dodenhoff, who has since put up more nesting boxes in her backyard, are attempting to entice the birds to their property. Others, like Lawrence Zeleny, one of the founders of the North American Bluebird Society, have erected dozens of bird houses to help the creatures return to their former haunts.

"Placed in the right environment, nesting boxes do attract bluebirds, and thus can help build up local populations," says Eugene Morton, a biologist at the Smithsonian Institution in Washington, D.C. Adds Cornell University researcher Charles Smith: "If nothing else, these people are drawing important attention to the plight of the bluebird."

A member of the thrush family, the bluebird ranges only in North America, where it is divided into three distinct species. In each case, the males are more colorful than the females. The male of the eastern species sports a bright blue back, blue tail feathers and a russett breast. It breeds from the Rocky Mountains to the Atlantic. The western bluebird, marked by brownish feathers on parts of its back, lives from the West Coast to the Rockies. The mountain bluebird, all blue except for a whitish belly, spreads as far north as Alaska. All three range into Mexico.

Of the three, the eastern bluebird has suffered the most in recent years as its rural farm habitat has given way to highways, shopping centers and subdivisions. The hollowed-out trees where it builds its nests have become scarcer, and the rotten wooden fenceposts it also inhabited have been replaced in many areas by metal.

To make matters worse, the natural nesting sites that remain are often taken over by the European starling or the house sparrow, two aggressive, hardy birds imported from England in the late 1800s. "The biggest problem bluebirds face is this competition over housing," says Zeleny. "These tough immigrants frequently usurp the bluebird nesting boxes, throwing out or attacking the adult bird. If the bluebirds are nesting, the starlings will sometimes drive them out and possibly kill their nestlings."

In the northern part of their range, most bluebirds migrate south in loose flocks during the winter to areas where their food supply of insects and berries is abundant. Others tend to remain throughout the year where they breed. During winter, the birds live in flocks. Some birders have seen more than a dozen of the creatures enter a single roosting box at once, and then huddle together through the night to keep warm. But in late winter, the birds turn to domestic affairs. After the birds pair off, they select a likely nesting site. During the next four to five days, the female builds a nest—usually a loosely constructed cup of grasses—while the male remains in the area.

Bluebirds usually produce two or three broods a year, laying three to eight eggs per clutch. Following a two-week incubation period completed solely by the female, the eggs usually hatch on the same day. Then for the adults, it's an almost continuous hunt for food, as their hungry youngsters require feeding about once every 20 minutes. After about 18 days, the young birds leave the nest, flying to a nearby tree or perch. Within a few days, the father takes over caring for the youngsters while the mother prepares the nest for the next brood. After the fledglings can fend for themselves, they sometimes stay around to help the parents feed the next brood.

With so many young growing so quickly, it's not surprising that bluebirds need plenty of nesting sites to flourish. "They're naturally curious," says Morton. "They like to investigate dark holes that might lead to a nesting cavity. Artificial nesting boxes provide a strong stimulus: a dark hole in light-colored wood."

People have been supplying such boxes for more than 150 years in this country. "In fact," notes Zeleny, "Thoreau's diary speaks of several bluebirds that nested in his boxes." Zeleny's own efforts span seven decades. "When I was about 14, I began trying to entice bluebirds into my homemade bird-

houses," he says. Today, he maintains a 60-box bluebird trail which he established 17 years ago near his home in Maryland. Such a trail is made up of strategically spaced, well-monitored nesting boxes with holes an inch-and-a-half in diameter to keep out starlings. Zeleny spends about five hours a week during the nesting season monitoring his boxes, keeping meticulous records of eggs hatched and birds fledged. He believes that about 2,500 young birds have fledged on his trail since it was first put up in 1967.

The bluebird society Zeleny helped found five years ago has since grown to more than 4,000 members. Camp Fire, Inc., a youth organization, has also taken to building nesting trails. In the past three years, some 10,000 children have been awarded badges for bluebird conservation. The group recently

Longtime bluebird advocate Lawrence Zeleny (above) checks one of the 60 nesting boxes he put up 17 years ago at a federal research facility in Maryland. Nearby, a female bluebird pokes her head into a box to feed her hungry five-day-old brood (left).

helped establish a 112-box trail at the Air Force Academy in Colorado. Last year, 46 were occupied by bluebirds.

The main thrust of the bluebird revival, though, has come from individuals, many of them retired people who grew up when bluebirds were abundant. Around Tupelo, Mississippi, for instance, people are seeing more of the birds these days, thanks to the efforts of 88-year-old Gale Carr, a retired dairy farmer. Twenty years ago, Carr was surveying a site for a new home when he spied a bluebird in a nearby tree. It was the first Carr had seen in some time, and it inspired him. From 1964 until a few years ago when he became disabled, Carr made more than 6,500 nesting boxes, which he gave away to people in Tupelo and some 20 other cities. "The boxes really attract the birds," says Carr. "Today, there are a great many bluebirds in this area."

Carr's northern counterpart may be Junius Birchard, a modern-day Johnny Appleseed from Hackettstown, New Jersey, who travels through parts of his state, dotting the landscape with nesting boxes and getting other people to do the same. For the past six years, the spry, 69-year-old former teacher has been visiting area schools and local clubs, preaching bluebird conservation.

"As a boy in Pennsylvania, I'd go camping along the Allegheny, and there'd be bluebirds everywhere," he says. In 1977, Birchard read an article about the eastern bluebird's decline. It spurred him to action. "At the time," he recalls, "I didn't even know where to begin to find a bluebird."

Before long, Birchard thought of a way to combine his love for teaching with his new mission to help bluebirds: "I arranged to teach a group of second-graders about bluebirds by helping

them build nesting boxes," he says. "It fit in beautifully with their curriculum, which called for using tools and also learning about various aspects of the environment." With Birchard's help, the students built some 50 nesting boxes and the following season, they recorded a successful nesting.

Word of his program spread quickly, and, he says, "the thing just mushroomed." As more and more groups began asking for his program, Birchard developed nesting box kits that people could easily assemble themselves. Originally, he cut the lumber for the kits himself, but soon enlisted the aid of students at two nearby schools. To date, Birchard has sold more than 6,000 of his nesting boxes. They sell for $3.00 each, and he turns all profits into the purchase of more lumber for boxes. So far, he's spent more than $10,000 of his own money on the project. Each of his

or later," he says, "we're going to be hit with a severe winter similar to those of the late 1970s, when large numbers of the birds died. In cold weather, food supplies are limited. Perhaps if the species were more widely dispersed, a larger percentage of the birds could get enough food to survive."

Ornithologist and renowned bird artist Roger Tory Peterson disagrees. Density, he says, is not a problem for bluebirds. "The birds will sort it out," observes Peterson. "They'll space themselves. When it gets too crowded, some of them will move to the periphery."

"Undoubtedly, when you entice birds into suburban situations, as many people are now doing, you increase the risks of mortality," adds Craig Tufts, chief naturalist with the National Wildlife Federation. "We're finding, though, that the number of birds is increasing far faster than the number that is being lost to predation and disease."

Can a nesting box program really help safeguard a species? "It's not a new phenomenon," says Eugene Morton. "The wood duck was nearly extinct in the 1930s. But through a similar citizen effort, it was protected with nest houses. Now, the wood duck is very common and no longer dependent on humans for its survival."

Though no accurate population figures are available for bluebirds today, scientists at the U.S. Fish and Wildlife Service Research Center in Laurel, Maryland, are beginning to see some trends. "Bluebirds appear to be increasing in some local areas," observes biologist Chandler Robbins.

Whether or not they ever come back in vast numbers, bluebirds are indeed returning. Their revival should be delightful, both for those people who remember the little songbird from years past, and for those who will see one for the first time. ∎

Eileen Lanzafama is assistant photo editor of this magazine. Maryland photographer Michael L. Smith has been attracting bluebirds to his yard for ten years. For more information about nesting boxes, write to the North American Bluebird Society, P.O. Box 6295, Silver Spring, Maryland 20906-0295.

bluebird boxes is stamped with an emphatic plea: "Please do not let the very destructive English sparrow breed in this birdhouse."

Another determined self-starter, Laurance Sawyer of Ringgold, Georgia, has been building and delivering nesting boxes to area residents since he retired from a local bakery in 1975. Sawyer fashions his boxes out of hollowed-out logs and sells them at a minimum fee. He produces nearly 1,000 boxes a year. "Many of the older people who want a box don't have the tools or the strength to put them up properly," he says. "Some of them simply need the therapeutic value provided by offering care to a living creature."

Jack Finch, a retired tobacco farmer from Bailey, North Carolina, channels his energy in another direction: he invents gadgets to keep predators away from the nests. After maintaining a bluebird trail of more than 2,000 boxes for several years, Finch realized that he couldn't take care of the project properly. He took down the boxes and replaced them with 360 new birdhouses perched atop ten-foot pipes. Finch, who had also invented a device to keep snakes out of the boxes, created the new design to stop flying squirrels from nesting in the bluebird homes. "It works," he says.

Despite such success, some scientists question whether it's wise to attempt to restore large numbers of bluebirds to small areas. T. David Pitts, a researcher at the University of Tennessee at Martin, suggests that by doing so, people may be inadvertently exposing the birds to some of the same problems that wiped out the creatures' populations in certain areas in the first place. "Sooner

He Migrates With The Seasons

Seasons

In his ceaseless studies of sanderlings, biologist Peter Myers travels the length of the Americas

By Nyoka Hawkins
Photographs By Frans Lanting

A T 2:00 A.M. ON a cold December night, six figures march through the darkness of Salmon Creek Beach near Bodega Bay, California. Wearing hip boots and headlamps, they carry several long poles, binoculars and plastic cages strapped onto backpacks. After a mile or so, they stop, place one pole on the beach and pull another into the water, with 60 feet of netting stretched between. They have spotted what looks like a piece of driftwood just below the lip of the sand. Slowly, quietly, they walk toward it. The wood starts to pulse; its borders seem to blur. They pause, then begin their approach again. The wood moves away. Again and again, for an hour perhaps, they draw near, until when the wood has moved into just the right position, the figures start to howl and shriek. Instantly, the "driftwood" explodes and a flock of 500 sanderlings take flight, about 50 of which fall directly into the carefully placed net.

The capture is the culmination of the evening's events, but for zoologist Peter Myers, who has led dozens of these expeditions, the work is far from over. Quickly, he and his associates, who work out of the University of California's Bodega Marine Laboratory, must get the sanderlings into the cages before the crashing waves carry the net and birds out to sea. "At least once a night someone fills his hip boots with water," says Myers, currently a curator at the Academy of Natural Sciences in Philadelphia. "But we go through it because once we get the different colored bands on, then we can tell the birds apart—tomorrow when we let them go, or next year when they return from migration. It allows us to study each bird as an individual."

For eight years, Myers, 33, has been studying the sanderling, an eight-inch puff of feathers that belongs to the sandpiper family. Making its winter home on beaches throughout the world, it chases the waves from August to late May, feeding on the tiny invertebrates left stranded in the sand, then begins its

long migration north to breed on the treeless plains of the Arctic tundra. Traveling at an altitude of more than 10,000 feet, and at speeds up to 50 miles per hour, the sanderling flies through the thin air, sometimes for 35 hours without stopping, until it reaches one of its staging areas—a temporary but critical feeding ground that enables the bird to continue on an annual voyage that can reach up to 8,000 miles.

Identifying and protecting these staging areas is one of the most important aspects of Myers' research. Over the years, the soft-spoken, determined researcher has changed the way other scientists think about shorebird behavior, particularly territorial defense. To discover more about sanderlings, he has traveled the length of the Americas. His work has taken him north to Barrow, Alaska, where, finding himself alone in the field with no way to make coffee, he once ate coffee bean and peanut butter sandwiches ("Disgusting," he says, "but it had the necessary effect"). And it has taken him south, to the far reaches of South America, where last fall, a jeepload of Peruvian marines descended on his crew of ornithologists while they were looking for the nest of a local bird. "They thought we were trying to blow up a power station," recalls Myers.

Originally, Myers' work concentrated on the Bodega Bay population of sanderlings, which returns to that site year after year and which numbers about 700 birds at the height of the winter season. His goal was to find out why the birds only defend their territories intermittently. Sometimes, the birds show signs of fierce territoriality, fighting each other incessantly to maintain exclusive rights to a 40-yard strip of beach; other times, they feed together in flocks, displaying no hint of aggression. Working with zoologists Frank Pitelka and Peter Connors, Myers has uncovered a number of different facts about the nature of the sanderling's defense and the reasons behind it.

"For a long time," he says, "researchers had noticed that there is an inverse relationship between the amount of food available and the size of the terri-

Strutting into the tide, a sanderling searches for food along the California coast (above). The tiny beachcombers inhabit many of the world's seashores.

Some sanderlings migrate as far as 8,000 miles twice each year, flocking together both in the air and while resting (opposite page). The birds often fly from wintering grounds in South America to breeding areas in the Arctic.

tory an animal defends. The more food, the smaller the territory. They explained this by saying the bird decides there is enough food in a given small area, so it doesn't need to defend a larger one. However, we found that, given the chance, a bird will defend as large an area as it possibly can. But, of course, more birds move into the good feeding areas, and it just takes too much energy to fight off all the intruders. We also learned that a bird will not defend a territory that has no food, or which has an overabundance of food."

The roots of Myers' work with sanderlings reach back to 1971 when, as an undergraduate traveling in South America, he came upon an island which Charles Darwin had researched. As Myers sat reading about Darwin's travels in *The Voyage of the H.M.S. Beagle*, and watching thousands of sooty shearwaters migrating north, he decided to study zoology. In 1973, after a year of graduate school at Berkeley, he re-

turned to South America with a fellowship to study shorebirds. For the next year and a half, he and his wife, Lois, traveled through Argentina. A shorebird novice when he first arrived, Myers says: "I knew so little, I discovered several new species the first week—because I was misidentifying everything. A professional ornithologist there was horrified."

With the help of grants from the World Wildlife Fund-U.S., Myers is expanding the scope of his sanderling study beyond Bodega Bay to South America, where the largest concentration of sanderlings in the Western Hemisphere is found along the coasts of Chile and Peru, feeding off the bounty of the rich Humboldt Current. "When I went to Peru last year," recalls Myers, "in addition to being chased by marines, I was staggered by the numbers of sanderlings on the coast there."

Currently, Myers is trying to discover why some sanderlings winter as far north as Bodega Bay and others fly on for thousands of miles to South America. "Based on research that we've done on the Peruvian and Chilean birds," he says, "life looks a lot easier down there. On average, the birds living along the Humboldt Current are heavier. They spend a lot less of their time feeding. The birds in Chile roost most of the day. At Bodega Bay, the sanderlings lose weight through the winter. It gets hard for them to find enough food. Their northern location, though, is closer to the nesting ground and that may affect their breeding success, which, of course, is the bottom line."

In addition to gathering biological information on the sanderling populations in Chile and Peru, Myers is training South American scientists in the techniques of studying and managing migratory birds. "Wetland habitats are disappearing at an alarming rate down there," he says, "largely because of agricultural development. Many Latin American countries simply cannot afford the luxury of serious conservation efforts. My goal, and that of the World Wildlife Fund in getting grass-roots efforts at research and conservation started, is to help bring these countries

Stopping during its travels to Latin America, a flock of sanderlings rests at a staging area. To reach one of these temporary feeding sites, the birds may fly up to 35 hours without a break.

to a point where they can assess for themselves the environmental costs of development. Right now, many of them are insensitive to the local impact of draining wetlands, much less the international implications."

Because shorebirds move over such a wide range, Myers hopes to establish an international network of management areas to aid in their conservation. "When you have species that are flying from the Arctic to the United States to South America and back again," he says, "they pose a real challenge to traditional conservation strategies. You can spend billions of dollars protecting some areas, and still have serious problems because nothing's being done at another key site in their migratory route. Right now, as far as we can tell, the populations of sanderlings and most other shorebirds are holding their own, but abundance doesn't mean they're immune to extinction. Shorebirds are particularly vulnerable to habitat manipulation. Although they are scattered throughout the world, they concentrate in a very few local sites that are crucial to their survival, especially in migration."

Myers notes, for example, that the en-

tire Atlantic Coast population of sanderlings from throughout the Americas may move through the Delaware Bay area in New Jersey for a three-week period every spring. This is in addition to a number of other shorebird species. "And now there's a suggestion that sanderlings from the South American Pacific coast move through there as well," he observes. "This spring, we spotted a bird we banded in Peru at Delaware Bay. That's a fascinating hint about the future. It means that a vast number of shorebirds, even more than we previously suspected, are very sensitive to what happens to the health of the Delaware Bay. Identifying and protecting sites like this is a very critical component of what we're doing. These birds need to stop at that resource en route to their breeding ground."

Those birds who do make it to the breeding ground find a burgeoning Arctic summer waiting for them. In the mossy expanse of sedge grass and dwarf willow, near pools of melted frost swarming with insect life, the female sanderling lays her clutch of four eggs, her nest no more than a scrape in the ground formed when the male sanderling presses his breast into the soft pas-

Stalking sanderlings in the surf, Myers plants a mist net on a California beach (above, left). His nighttime efforts are rewarded when he later returns to tag and release the birds (above).

The scientist has discovered that the shorebirds' feeding behavior depends on the supply of food. Often, sanderlings flock together peacefully, like these birds (right). But when less food is available, they become very territorial.

ture and then lines the shallow indentation with lichen and dried grasses.

In less than four weeks, the first tiny chick will hatch, and then dry itself for an hour or so in the wind and sun. At the seasoned age of 30 days, long after the parents have departed, the bird begins its first journey south. To Peru or Chile. Australia or Patagonia. Or to Bodega Bay, where one night by the dark rolling surf, it might fly into a net and help a group of screaming, soggy researchers advance the boundaries of zoological science. ∎

Philadelphia writer Nyoka Hawkins interviewed Peter Myers at the Philadelphia Academy of Natural Sciences. Dutch-born photographer Frans Lanting lives in California.

PHOTOGRAPHS BY LEONARD LEE RUE III

Something to Gobble about

Wild turkeys are thriving throughout the country ...a fitting status for such a celebrated bird

BY RENIRA PACHUTA

ACCORDING to the dictionary, a turkey is not only a bird, but also "a failure, a flop or an incompetent." What's the connection? That's a bit of a mystery. "There's nothing incompetent about a wild turkey," asserts Gene Smith, an official with the National Wild Turkey Federation headquartered in South Carolina. "It is one of the craftiest, most elusive creatures around. To outsmart one, you have to 'think' like a turkey, and you have to act like one. Not only is its hearing better than yours, but its eyesight is also much keener. And what's more, it can run a heck of a lot faster than you. A sportsman would be hard-pressed these days to find a more challenging game bird." That hasn't always been the case.

Long before the Pilgrims landed at Plymouth Rock, scientists believe, at least ten million wild turkeys ranged throughout North America. Back in those days, the birds apparently showed little wariness toward humans, and thus they were easy targets for Indians. Among many native tribes, in fact, it was considered beneath the dignity of the adults to hunt the birds; the task was usually assigned to children.

If the birds gradually became wary of people, it was because they had to, in order to survive. The reason: as European colonists expanded their settlements across the continent, the turkey's woodland habitat increasingly disappeared — and with it, so did the birds. By the late nineteenth century, the creature had been eliminated from much of its original 39-state range, causing one naturalist to write that the wild turkey would soon be "as extinct as the dodo." Fortunately, this gloomy prediction never came to pass. Although, according to one estimate, only about 30,000 wild turkeys survived throughout the United States by the end of World War II, today the birds are thriving better in their natural habitat than at any other time in this century. "It's one of the most dramatic wildlife management success stories in recent history," says Smith. Some two million turkeys now inhabit forests in every state except Alaska, and spring gobbler hunts are conducted in 43 of those states.

Despite its name, the turkey has nothing at all to do with the western Asian country. It is a uniquely North American game bird, and six subspecies roam the continent. The most common is the eastern turkey — the bird featured on these pages — which ranges from the Midwest to the Atlantic Coast. At the first Thanksgiving celebration in 1621, local Indians donated "great stores of wild Turkies" to the Pilgrims as their share of the feast.

These days, however, most of us dine on turkeys that are descendants of creatures once domesticated by the Aztec Indians of Mexico. Those dumpier, shorter legged cousins of the wild turkey were introduced into Europe in the early 1500s by Hernando Cortes as "one of the spoils of war." Eventually, European colonists brought domesticated turkeys back to the New World.

Some years ago, in an effort to restore turkey numbers, biologists began breeding wild toms with domestic hens. The resulting offspring, however, displayed little instinct for survival in the woods. Many of them sought out the nearest barnyard and mingled with domestic fowl. Others became easy targets for poachers and predators. As a result, researchers turned their attention toward capturing wild turkeys and relocating them into new areas. The method has continued to work — though not without perseverance on the part of the scientists.

"Last winter," recalls Bob Eriksen, a

A feast for the eyes, a gobbler like this male eastern turkey is more often heard than seen. Smaller than its domesticated cousin and one of North America's most elusive birds, the wild turkey was once nearly eliminated from its original 39-state range. But recent wildlife management efforts have successfully reintroduced the birds into every state except Alaska.

A gobbler attracts a pair of smaller hens (left) in early spring by strutting through a clearing and bellowing his distinctive call. During courtship, the male puts on his finest colors. His iridescent tail feathers fan out (above), his chest puffs and his fleshy wattles fill with blood and turn bright red.

biologist with the New Jersey Division of Fish, Game and Wildlife, "we waited from dawn until midafternoon, three days in a row, hoping to capture a flock of hens and their young for relocation. But each day, the turkeys were scared off by cross-country skiers and snowmobilers before we could try to capture them. Then, on the fourth day, as the birds approached the feeding site we'd set up and we were about to net them, a team of sled dogs suddenly came tearing through the woods. We gave up after that." Despite such minor setbacks, says Eriksen, "the situation for turkeys is definitely healthier."

While few people are lucky enough to see a wild turkey, many do hear males gobbling for their mates in springtime. If a tom is in good voice, he'll be joined by one or more hens, and the courtship display begins. The tom's head and neck, which are practically bare of feathers, suffuse with color as the blood rushes through, turning from red to blue to purple. Strutting up and down in front of his future harem, a tom drops his wings, fans open his tail feathers and puffs up his chest, giving the impression of a large beach ball. Then, making a puffing sound, he'll de-

105

After mating, a hen builds a simple ground nest and incubates her clutch of speckled eggs (above) by herself. After about a month, young turkeys emerge and begin feeding themselves.

flate and step forward a few more paces, dragging stiff wing feathers across the ground. No wonder the wing ends are worn down by the end of the season — and, after many such performances, so are the turkeys.

After mating, the hen takes on the responsibilities of a "single parent." Choosing a well-camouflaged nesting site in or near a wooded area, she lays about a dozen eggs and incubates them for the next 28 days until they hatch. By late fall, when the poults are several months old, the young males are chased away by the hen and form their own small groups. The females, on the other hand, stay with their mother until they reach maturity.

Though many wild turkeys manage to avoid winding up on the dinner table, they don't necessarily live to a ripe old age. "The record that we know of for a tagged bird is 12 years," says Gene Smith, adding that habitat preservation is very crucial to the birds' well-being in the years ahead. "They are one of our unique wildlife heritages," he observes. So, the next time someone calls you "a turkey," you might thank him for the compliment. ∎

Renira Pachuta is an assistant editor of this magazine.

What sounds like a locomotive, sports a bizarre beak and seals itself up to tend its eggs?

HORNBILL

A most peculiar bird

BY PAUL SPENCER WACHTEL

IT sounded as if a runaway steam engine containing a crew of yelping puppies was closing fast on our longboat on Sarawak's Apoh River in northwestern Borneo. "What's that?" I asked my friend, Nawan, a Kayan tribesman. "Just a hornbill," he replied. "No reason for us to stop." That's how I found out that the bird is one of the noisiest creatures in the tropical rain forests of Southeast Asia. However, that bit of knowledge did little to prepare me for my first look at this startling creature several days later.

The hornbill was a pet which was tied to a post in a village longhouse. It was as big as a goose and black as a starling. It had the eyelashes of a debutante, a skinny neck like that of a buzzard with a goiter and a long, humped beak straight out of a Ringling Brothers' sideshow. I knew that, in spite of the bird's physical appearance and eccentric call, I was looking at one of the more sought-after and revered creatures of Asia.

The beak is the hornbill's most distinctive attribute. Superficially, it resembles that of a New World toucan, though the birds are not related. There are at least 44 species of hornbills, which range from Africa to the Solomon Islands, and many of them sport a peculiar protrusion atop the beak. It's called a casque, and in some forms, it sticks up like a 1959 DeSoto tail fin with an overdose of hormones. Much lighter in weight than it looks, it is a horny shell supported internally by spongelike tissue.

What purpose does the casque serve? That's a matter of considerable speculation. Some scientists think it is a means of attracting the opposite sex much the same as other birds do with brightly colored plumage. Other researchers say that the casque helps the large bird dissipate heat. In most hornbill species, the inside contains a network of tissue containing numerous blood vessels which rapidly release heat, keeping the bird comfortable in the sweltering tropical rain forests, which many species inhabit. The helmeted hornbill's protuberance is the heaviest and may have other uses. "I think the helmeted hornbill's casque might also be used as a weapon," says Ken Scriven, director of the World Wildlife Fund in Malaysia. "I found a helmeted hornbill that had apparently been killed in a fight. It's possible the birds wield their heavy casques like truncheons in a head-to-head battle." The thing certainly *looks* formidable enough!

The helmeted hornbill is the only hornbill with a solid casque, and this feature has caused it no end of trouble.

Malays of Perak State make poison-detecting spoons out of its casque, the idea being that the spoon will cause a lethal brew to bubble and simmer. Across the Malay Peninsula, traditional healers of Kelantan State mix the casque with elephant tusk, dugong bone and white hibiscus to make an antidote for cyanide of potassium poisoning. The casque of this magnificent four-foot bird also has long been sought by Chinese artisans who carve the precious substance called *ho-ting* into belt buckles, plume holders, decorative miniatures and snuff bottles. Centuries ago, Chinese traders traveled to Borneo and exchanged glass beads and iron tools for the golden orange casque, which was valued more highly than jade. Like all hornbills, this species is still intensively hunted wherever it is found, and its prospects are closely linked to the future of the tropical rain forests — which are being cut and burned at a rapid rate.

While the Chinese were trading trinkets for the helmeted hornbill, the Ibans

A bump on the beak — known as a casque — is a trademark of most hornbill species. Prized by carvers and folk healers, it varies from the warty blob of the Blyth's hornbill (right) to the streamlined version of the Indian pied hornbill (overleaf). The birds' behavior is as strange as the casques.

Looking like a chicken with a hangover, this white-crested hornbill is one of the few species with almost no casque at all.

of Borneo were worshiping its relative, the rhinoceros hornbill. This is the sacred Iban bird of war, the *kenyalang*, which is perhaps the most glorious hornbill of them all with its huge, upturned casque the size of a banana. Birds are important omens throughout Borneo, but this impressive creature, which is one of the eight hornbill species found on the island, is the only one that has a festival named after it: the *gawai kenyalang.*

I was invited by an Iban friend to his longhouse to take part in one of these celebrations, a five-day blast which once was held to honor a returning headhunter. "Headhunting is no more, of course," my friend explained, "and now we usually hold the festival to celebrate a really big occasion, as when our paramount chief was elected to the Malaysian Parliament."

When dinner was served, we proceeded to stuff ourselves with wild pig, dog meat, chicken and a potent beverage called *arak*. Out of the corner of a glazed eye, I saw the party's principal

totem — a rhinoceros hornbill icon, which represents *Singalong Burong*, the Dayak God of War. Although modern Ibans are well-educated and frequently Christian, they still respect the old traditions.

Dayak warriors wear hornbill feathers in their caps and war coats, and proudly display earrings of helmeted hornbill casque carved in the shape of a leopard's canine tooth. In olden times, before a battle, large wooden rhinoceros hornbill images were used to send the statue's soul against the enemy. These are sometimes faddish masterpieces. J.D. Freeman described a carved wooden hornbill he saw several years after World War II. It was, he said, "decorated with a Sarawak ranger astride an elephant, pointing his rifle threateningly at a Japanese soldier, and so protecting a nomadic Punan . . . ."

The hornbill is at its noisiest when flying; it has air spaces at the bases of the primary wing quills, which produce a bellowslike sound while flapping through the air. All hornbill species have distinctive calls, but the most interesting is that of the male helmeted hornbill. Local myth suggests that its call replicates the sound of an angry son-in-

law chopping down his father-in-law's house. It starts softly with a series of "tooks," builds in tempo and volume over ten minutes or more, and ends in an amplified climax of frenzied laughter which defies description.

If you were to follow the call of a hornbill, you might notice orangutans, the great apes of Borneo and Sumatra, doing the same. Jan Wind, an ornithologist with the World Wildlife Fund in Bogor, Indonesia, says that hornbills and orangutans both favor large fruits, particularly figs, and often the red apes follow flocks of up to 50 of the big birds in search of fruiting trees. But fruit isn't the only component of the hornbill's menu. Some species, particularly the ground hornbills of Africa, enjoy insects, lizards and snakes, along with an occasional bird.

Noisy as the hornbill is in its daily activities, it draws a curtain of privacy around its unusual nesting habits. A naturalist friend at Jakarta's Ragunan Zoo told me how, in many species, the female imprisons herself during the four- to six-week incubation period and subsequent six-week rearing period. She first seeks out a large, hollow tree, dropping sticks and debris into the cav-

Jack Fields (Photo Researchers)

Tom McHugh (Photo Researchers)

As agile as a kid tossing popcorn, a hornbill downs some fruit with a fillip of its ponderous saw-toothed beak.

ity to raise its floor to a comfortable level. The female hornbill then spends several days plastering up the entrance with her own droppings, which harden into an impenetrable claylike substance. The male helps from the outside, using the flattened sides of his bill like a trowel until only a narrow slit remains.

Thus protected from predators, the female will stay in this cell while her mate feeds her fruits. One observer recorded 18 visits by the male in ten hours, and another bird-watcher, no doubt bleary-eyed and semicatatonic from looking up a tree for 3½ months, recorded that one male brought 24,000 fruits to his mate during that time. A male hornbill takes his responsibility seriously. One bird was seen repeatedly bringing his imprisoned consort flowers, which are not part of the hornbill diet, but certainly add a nice touch.

To catch a female hornbill and her young, all you need to do is find a nest and haul them out, since the female quickly molts and becomes flightless

and very fat during part of her imprisonment. Wallace, the British naturalist who, together with Darwin, helped formulate the theory of evolution, wrote of coming across a hornbill nest in Sumatra: "After a while, we heard the harsh cry of a bird inside.... I offered a rupee to anyone who would go up and get out the bird...but they all declared it was too difficult, and they were afraid to try.... In about an hour afterward, much to my surprise, a tremendous loud, hoarse screaming was heard and the bird was brought to me, together with a young one which had been found in the hole. This was a most curious object, as large as a pigeon, but without a particle of plumage on any part of it. It was exceedingly plump and soft, and with a semitransparent skin, so that it looked more like a bag of jelly, with head and feet stuck on, than like a real bird."

The adults are equally intriguing. One afternoon at a zoo, I was entranced by a rufous hornbill from the Philippines. It hopped around with both feet leaving the ground at the same time. Although it is dangerous to anthropomorphize, I got the feeling that the bird knew just how silly it looked, and tried

The rhinoceros hornbill is a war-god in Borneo. As its tropical forest habitat dwindles, it needs all the help it can get.

to make *me* feel bad by staring me down with its big eyes.

Suddenly, it made an astonishingly delicate maneuver. It bent down to pick up a berry, but it couldn't simply swallow the fruit, since, as in all hornbills, its tongue is shorter than its beak. It had to flip up its food like a piece of popcorn and gulp it down. Then the bird made a most unusual burp — a cross between the scream of a gagged mezzo-soprano in distress and the not-very-polite sound of an obese diner in urgent need of Alka-Seltzer.

I gained renewed respect for a bird that could get away with a repugnant sound like that in front of the small children who stood in awe next to its cage. One little girl, her eyes wide with delight, remarked: "Mommy, look. After God made all the pretty birds, he must have gotten tired and made this one." ∎

Paul Spencer Wachtel is public affairs coordinator for the World Wildlife Fund in Switzerland.

113

Armed with an arsenal of natural weapons, the great horned owl is well equipped to pursue and catch everything from a robin (right) to a rattlesnake.

HOMING IN
on the
HUNTER

*Scientists are unlocking the secrets behind
the great horned owl's predatory prowess*

By David Seideman

ERE she comes," Leroy Petersen murmured to himself. No sooner had the Wisconsin Department of Natural Resources biologist reached the great horned owl's nest at the top of a tree, than the mother slammed into him from behind, knocking a baseball cap off his head. The incident reinforced Petersen's respect for the owl's legendary fierceness. Indeed, few birds are so well equipped to ward off enemies, or to pursue and capture their prey.

The great horned owl's arsenal includes a wide assortment of weapons. Some researchers believe it may have the most acute sense of hearing in the animal world. Short, feathered flaps funnel sound to the deep slits on the side of its head—where the bird's ears are located rather than, as one might expect, inside the horn-shaped tufts on top. These ears can pinpoint high-pitched notes, such as the squeak of a mouse or the faint sound of a beetle rustling through the grass, more than 100 yards away.

The owl does not rely solely on its ears to locate prey, however. Though the bird weighs only about three pounds—which still makes it one of the largest owls native to North America—its eyes are as large as a human's and 35 times more sensitive. The owl's vision is so keen that it can closely monitor prey far off on the horizon—prey that would be undetectable to the human eye. The raptor's eyes also admit a remarkable

> # *When I approached a nest, an owl suddenly dive-bombed me. I felt like I was being hit by a sharp brick."*

Quick learners? These nestlings will soon test their flying ability. Most owlets attempt inaugural flights about three months after they hatch. More often than not, the result is a crash landing.

Though it weighs only three pounds, this horned owl (right) has eyes that are as large as a human's—and that are 35 times more sensitive. The raptor can monitor prey far off on the horizon.

amount of light. As a result, the horned owl can sight an object with roughly five percent the intensity of light that a human requires. Moreover, the bird's swiveling, mobile head enables the animal to sight prey in almost any direction without moving its body.

Taking off from its perch at the edge of a clearing, the owl homes in on a vulnerable target. The shapes of the bird's wings and feathers reduce flight noise and thus help maintain the element of surprise. At the moment of attack, the assailant swoops down, tosses its head back and legs forward in one rapid motion, and kills most prey instantly by sinking sharp talons into the body. Victims range from rodents and rabbits to rattlesnakes. "The bird is very opportunistic," explains Bruce Fortman, a biologist at East Stroudsburg University

in Pennsylvania. "It will take whatever is available."

The "winged tiger," as the owl is sometimes called, rarely backs off from a challenger when defending its nest. Fortman learned this firsthand. While studying the bird, he once had to beat a hasty retreat after being attacked by an apparently disturbed adult horned owl. "When I approached a nest," says Fortman, "an owl suddenly dive-bombed me. I felt like I was being hit by a sharp brick." Fortman's clothes were torn and his neck was badly cut. The researcher now knows not to venture into the field to study horned owls without wearing a helmet and heavy jacket.

The horned owl isn't always an adept predator, however. Learning how to fly, for example, poses enormous difficulties for any young bird. During the first

The horned owl's hearing is so acute it can pinpoint the location of some prey more than 100 yards away

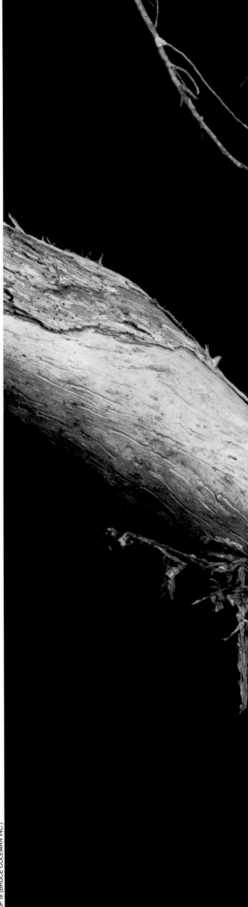

Guess what's coming for dinner? In southern Arizona, an adult owl brings a packrat back to its cactus nest (above). Like other birds of prey, young owls need parental care for several months.

Horned owls sometimes hunt by day, but their vision is better suited for nighttime activities. This bird (right) can sight objects with about five percent the intensity of light a human requires.

several weeks after hatching in February and March, horned owl nestlings rely on their parents for a steady supply of food. But they cannot live off their parents indefinitely. A parent may have to go to great lengths, and possibly withhold food, to force its offspring to attain self-sufficiency. Fortman has observed owls, perched on adjacent trees, waving food at fledglings to lure them into taking to the air.

By the beginning of the third month, most owlets attempt their inaugural flights, which, more often than not, end in crash landings. "Sooner or later," says Petersen, "a young horned owl will end up on the ground." Upon flopping furiously to the earth, the little bird struggles back up the tree. Gripping the bark on a trunk with its talons and beak, it climbs to a branch and tries again.

Once fully airborne during the next few months, the novice may overtake and catch prey and play with a potential meal rather than eat it. Before long, though, the juvenile instinctively begins to master the art of aerial hunting and fend for itself. In late autumn, the youngsters prepare to establish their own hunting grounds.

Until recently, many researchers overlooked or neglected the horned owl. Bruce Fortman wonders why. "We'd be hard-pressed to find a more efficient or adaptable predator," he argues. While other species have suffered the effects of pesticides and habitat loss throughout the Americas, the horned owl is holding its own. ∎

David Seideman is an assistant editor of this magazine.

C. ALLAN MORGAN

FRED J. ALSOP III (BRUCE COLEMAN INC.)

Courting With Gold

f nests were any indication, the bowerbirds of Australia and Papua New Guinea would be considered poor builders. The flimsy structures—built by the females—are little more than a few sticks. But the birds' courting areas, or bowers—built by the males—are architectural wonders. Male golden bowerbirds (photo) encase tree trunks and a horizontal display perch in thousands of short sticks covered with pale green lichens. For all this work, however, nobody has observed golden bowerbirds actually mating in the bower.

PHOTOGRAPH BY HANS AND JUDY BESTE

BODY SNATCHER

PHOTOGRAPHS BY REINHARD KÜNKEL

THE BEAK of the Rüppell's griffon is broad and powerful, hooked at the end for stripping flesh from the dead. Its ratchetlike tongue is large and gutter-shaped, with rows of long, horny "teeth" pointing inward and backward, presumably to help in shoveling down lumps of meat. Its neck is long and snakelike for reaching deep into an animal's body cavity to devour soft remains from the inside, and the neck itself is bare to repel gore that would stick to feathers. Unsavory as all that might seem, the big vulture is a natural—and vital—component of nature's scheme. Indeed, on the Serengeti Plain of East Africa, where the accompanying photographs were taken, it is an integral part of both dying and living, and as one of nature's undertakers, it is marvelously equipped for its mission on earth.

Though the Rüppell's "ghoulish" behavior at a kill is its hallmark, its behavior in the sky as it searches for the dead is in many ways even more distinctive. The bird is a stocky creature, weighing about 15 pounds, and although it is a lumbering flyer, its soaring abilities are extraordinary. Gliding on hot, rising air currents, it can travel as far as 100 miles a day from cliffside nests to its feeding ground, and it often stays in the air for up to six or seven hours a day.

Where there is even a hint of wind as the vast savannah heats up in the mornings, the vulture can hitch a ride on a spiral of upward-climbing air. It can cover almost three miles from a standing start in only six minutes. Aloft on these thermals, the sharp-eyed cousin of eagles and hawks watches for predators, especially the big cats, while carefully monitoring movements of other vultures. The birds will follow a cheetah on the hunt for hours, hoping to cash in on a meal at the end of their wait. When one vulture spies a carcass and drops to the ground, others follow in a crowd, sometimes to square off with hyenas and other predatory mammals for a share of the meal.

At the kill, the griffons frequently split the spoils with other flying scavengers. Altogether, there are six vulture species in the Serengeti National Park in Tanzania. All have a slightly different feeding niche, and all six may sometimes converge on a single carcass. White-headed vultures, followed by the lappet-faced, tend to arrive first. The lappet-faced rip open tough carcasses enabling Rüppell's and white-backed vultures, next in line, to gorge on the softer inside parts. Hooded and Egyptian vultures, both smaller birds, usually arrive last to glean any remaining tidbits.

A gathering of Rüppell's at a carcass is punctuated by many short, sharp outbursts. One bird will face down an opponent by stretching its neck obliquely forward and upward, offering a loud hiss—a threat display accompanied by partially spread wings and leaps into the air, all of which usually serves to change confrontation into accommodation. As the feeding progresses, a bird might even crawl inside a corpse for the juiciest morsels of rotting meat.

When the vultures get airborne again, they soar back to their cliffside nests or to treetop perches where they may roost in large numbers. Their breeding behavior may help explain the evolutionary pressures that make them so efficient in the air: for eons, successive generations had to commute long distances for food to feed their young; the best gliders were most successful in perpetuating themselves.

In the eastern Serengeti, there is only one major rocky outcrop—the Gol Cliffs—where the Rüppell's can nest. When the birds are rearing nestlings and need especially large amounts of nourishment, they can feed on the huge armies of wildebeest which calve nearby on the grassland plains. But when the young wildebeest are big enough to migrate and the great herds head for the horizon, the Rüppell's must follow. Dozens of them soar 60 miles, often more, in search of stragglers that will die on the trail, then return 60 miles to feed their chicks.

This vision of parental solicitude may not be enough to resurrect the image of a bird so distasteful in human terms. But in its dark, funereal dress, the Rüppell's griffon is at least as diversified and specialized in its lifestyle as the gorgeous flamingo. It takes the eye of an informed beholder to see the beauty lurking there.—*Norman Myers.* ∎

The Rüppell's griffon is one of Africa's most efficient undertakers

Gore dribbles from its beak as a Rüppell's griffon gorges on carrion (left) in a feeding frenzy in the Serengeti National Park in Tanzania. The vulture's neck is long and free of feathers, enabling the bird to reach inside a carcass to pull out soft flesh without soiling its plumage. Vultures monitor each other's movements as they soar across the grasslands, and when one drops to the ground, others follow in a crowd (below) to strip a body clean in minutes. The big birds are actually quite clean and bathe frequently.

High up in a cliffside nest (left), a Rüppell's guards its lone egg. Though stocky, the bird is a master of the savannah skies and may soar 100 miles from the nest for food (above). Run-ins with wild dogs (below) and other predators are common at a kill.

Dancing to a Different Beat

Almost everything about a blue-footed booby is comic—except to another booby

STORY AND PHOTOGRAPHS BY TUI DE ROY MOORE

I T'S a warm and sunny morning as I climb the island's slope, lured by loud honking and whistling. When I crest a rise, a familiar scene unfolds: several hundred pairs of seabirds are dancing noisily in an animated courtship display.

Small males, their stiff tails cocked, strut about to woo the larger females. Walking bolt upright, they "sky point," their bill, tail and wings stretched heavenward as the owners utter long, eager whistles. When the female joins in, the pace picks up. Staying close together, the two birds bow, touch bills, sky point in unison, and goose step, lifting their oversized feet rhythmically as if to the sound of a beat.

The performers are blue-footed boobies, and the whole peculiar ritual is carried out in slow, deliberate and, to my eye, comic gestures. In human usage, a booby is an awkward person—a dope—and, indeed, the birds earned their name because they seemed so clumsy. I cannot help but smile as I watch. Yet I know this elaborate behavior is funny only as a projection of my own sense of humor. To a booby, these antics are deadly serious. Without them, the species could not survive.

The blue-footed booby is one of six booby species and is closely related to the gannets of cooler climes. Distinguished, not surprisingly, by its bright blue feet, it ranges along the Pacific Coast of North and South America, where it nests in colonies, primarily on islands. Here, the species pirouettes and prances through a life that, for all its apparent comedy, can be downright grim at times.

Just finding food is a considerable problem in this unforgiving environment. The booby's main food is fish, and the supply varies greatly. An abundance of food usually triggers the beginning of a nesting cycle, but should the fish disperse during the breeding period, the birds may find themselves unable to travel far enough to shuttle food long distances back to their nests.

At one colony I visited, only dried

The mating rite of the blue-footed booby (left) is replete with sky pointing, bill touching and rhythmic goose stepping— behavior critical to booby survival.

A teeming booby colony (right) thrives on a narrow margin; if food sources move, the adult birds follow, even if it means the helpless chicks will starve to death.

carcasses and matted, stained down of starved chicks remained. Off to one side a half-dozen large juveniles, almost ready to fledge, sat morosely. When at last a solitary adult arrived from the sea with some food, this desperate mob assaulted it with such ferocity that the panicked parent fled without having located or fed its own offspring.

Though I was shaken by this sight, I realized that this natural disaster helped to select the fittest for survival and keep the birds' numbers in balance with their habitat. Natural selection may also be the force behind the ritualized movements of booby behavior. The silly-looking courtship serves to ac-

quaint the mates with each other, cementing the bond provided by their nesting instinct, and building up the trust and coordination that will be necessary to hatch and raise their young. Likewise, the clumsy motions of the young are merely first steps in learning the difficult art of acquiring food.

When the chicks hatch, they are gray little bundles, naked, almost blind, and totally helpless. However, baby boobies grow fast, and as their feathers grow, they spend more and more time flexing their wings, sometimes lifting an inch or two off the ground.

The booby's first flight starts off hesitantly, but the bird soon gains confi-

dence and begins diving into the ocean. When at last it is on its own, the grown chick will have begun to hone its survival skills. Even so, its future can be precarious.

The booby's fishing technique, for instance, is highly specialized, requiring a lot of energy and extreme precision to be successful. Typically, the bird cruises between 50 and 100 feet over the water, until it spots a fish, then swerves into a vertical plunge, flapping its wings powerfully to increase its speed of descent. At the last second before hitting the water, it adjusts its aim, then arranges its wings in a streamlined elongation of its body, piercing the surface like an arrow. A few times I have waded out to where a booby has been diving and, to my amazement, found that the bird had flung itself headfirst into water that was only two or three feet deep.

It is not at all surprising that young blue-footed boobies have a hard time mastering such behavior. Many die during their apprenticeship. Then the strange behavior of one of nature's natural comics is anything but a laughing matter.

Tui De Roy Moore is a naturalist who lives in the Galapagos Islands. She is the author of Galapagos: Islands Lost in Time *(Viking, © 1980).*

Growing up in a hurry

During its first weeks of life, a young booby is helpless and requires constant attention from parents (below), but even before its covering of white baby down is replaced by a new crop of feathers, "tiny" junior has grown big enough to squawk back (right).

MAMMALS

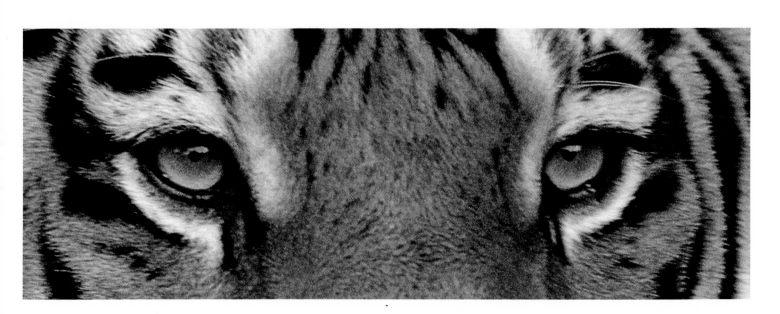

The rhinoceros is a docile creature most of the time, as it can well afford to be. For it is formidably armed. Perched on its nose is a weapon—a pair of horns—with which it defends itself against intruders. In addition to protecting the rhinoceros, these horns have, since the Stone Age, had a mystical significance for humans. Some oriental people have long believed that powder made from the rhino horn was useful as medicine and effective as an aphrodisiac, and that a cup made from one would purify poison in a drink.

Like the rhinoceros horn, the horns of other animals have held a fascination for people, and have been used for a number of purposes —mystical, practical and ornamental—that their original owners never quite intended. Thus, rabbis announce the observance of the Jewish New Year by blowing the shofar, a horn of a ram, in the streets of Jerusalem and in temples around the world. Indians of British Columbia make large ladles by steaming the horns of bighorn sheep over a fire until they crack open into two halves. In 1804, when Lewis and Clark blazed a trail from St. Louis to the coast of Oregon, they carried their gunpowder packed in bison horns. And big game hunters mount animal horns as trophies in their homes.

Horns themselves are relatively simple affairs—cones com-

ARMAMENT

...wners different ways

posed of a substance called keratin, the same material that makes up human fingernails, the talons and beaks of eagles, the hooves of horses, and even tortoises' shells. Unlike the branched antlers of the deer family, which drop off each year and are replaced, horns are a lifelong and continuously growing part of the animal. They are found exclusively on hoofed, grazing mammals. Because these animals are herbivorous, their impressive weapons are not used for getting prey. They are used by battling rival males in rutting season, and also as a defense against predators. The nature of this defense is one factor that determines the shape of horns. Some large African species, for instance, have horns that sweep backwards from their heads, providing protection from lions, which traditionally attack them by leaping onto their necks. Still, scientists cannot fully explain the amazing variety of horn shapes and styles, the unusual twists and spirals and curves that make each species' horn a unique design, useful to the animal on which it grows, and, for the rest of us, an object of beauty and fascination.

—*Stephen Freligh*

Like giant corkscrews, the horns of a cabul markhor spiral in classic grace. A wild goat, the animal (inset) lives in Afghanistan.

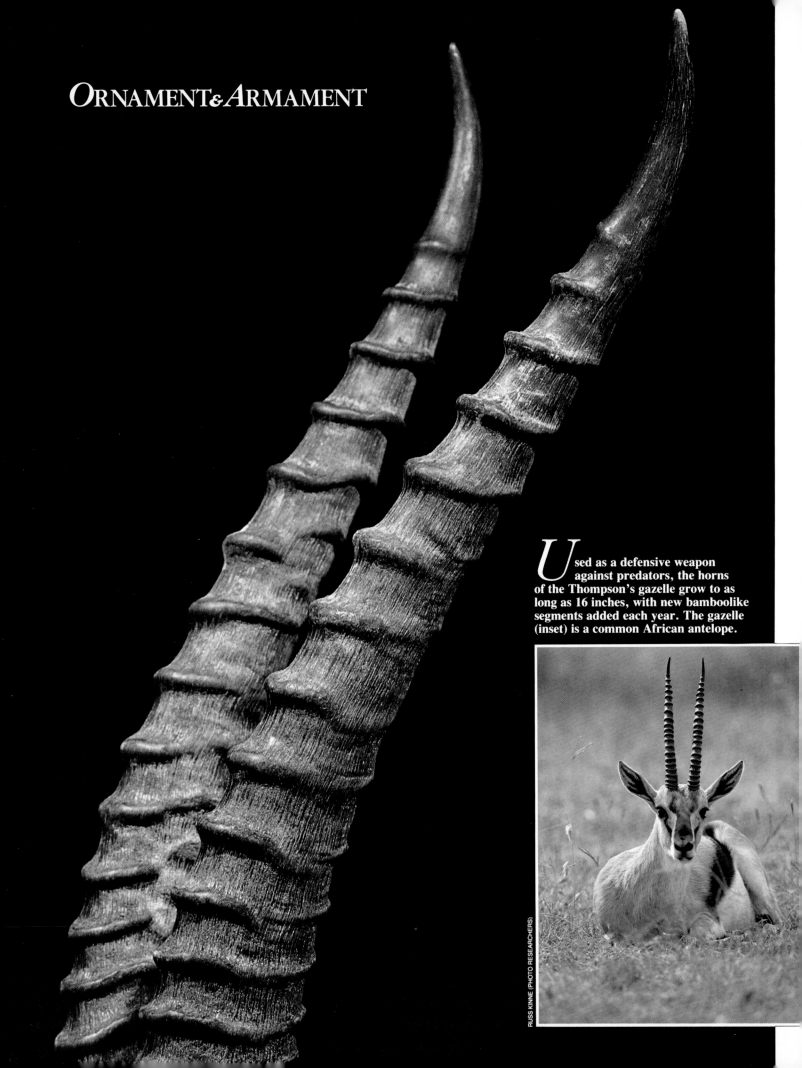

ORNAMENT & ARMAMENT

*U*sed as a defensive weapon against predators, the horns of the Thompson's gazelle grow to as long as 16 inches, with new bamboolike segments added each year. The gazelle (inset) is a common African antelope.

RUSS KINNE (PHOTO RESEARCHERS)

*F*ormed of hairlike fibers grown together, the rhino's horn is prized in oriental cultures for its alleged medicinal powers. The black rhino (inset) lives in Africa. All told, there are five rhino species.

SUSAN McCARTNEY (PHOTO RESEARCHERS)

ORNAMENT&ARMAMENT

*S*weeping skin and bone, the horns of a bighorn sheep can weigh ten pounds. Rams like this one in the Canadian Rockies (inset) use the bulky headgear to batter one another as they compete for ewes during rutting season.

Charging through the seasons

When it comes to year-round survival, moose aren't as clumsy as they look

WITH its small eyes, pendulous muzzle and heavy body perched atop coltlike legs, a moose looks like anything but a deer. Yet that's exactly what it is — the world's largest, in fact. Ungainly though the beast may look, it is almost as fleet-footed as its smaller relatives. And in open water, it swims so well that, at times, it can even outdistance canoeists. A moose is equally at home in alpine meadows, dense forests, frozen tundra and sprawling marshland. Further, as the accompanying photos show, it is admirably adjusted to the brief springs, short summers and harsh winters of the northern latitudes in which it lives — from Maine to Washington, from Wyoming to Alaska. — *Renira Pachuta*

Johnny Johnson

Spring
Calves are born and antlers bud

In moose country, the short spring always has an air of urgency. Last year's calves are chased away by their mothers and new ones are born. Triplets, such as these (left), are a rarity. Adult bulls shed their shovellike antlers each winter and begin growing new ones in April, when velvety knobs appear. The moose are so hungry that they bend down and scoot forward on their knees to get at the emerging vegetation.

141

Summer

The urge to eat is overwhelming

During warm months, the animals feed voraciously on aquatic plants. Calves, like these butting heads (above), grow from 30 pounds at birth to 150 pounds or more in a single season. A half-ton bull (opposite) needs 50 to 60 pounds of food each day to maintain its bulk and also grow four feet or so of antlers. Often, the urge to eat overcomes fear of humans; to a hungry moose, property lines and signs of human habitation (left) are meaningless.

Left: Martin W. Grosnick; above: Steven C. Kaufman; below: Michael S. Quinton

Autumn

Now, the bulls turn mean

After the growing season, the bulls begin their spectacular rutting displays. First, the animals clean and polish their antlers by rubbing off the soft velvety covering (left). Then, the interminable fights commence. Rutting bulls have been known to charge cars and trains. Battles like the one shown here (above) seldom cause serious injury. The victor (right) curls his lips and bugles — a summons to cows in the vicinity and a challenge to any males within earshot.

Above: Rollie Ostermick; below: Tom Bledsoe; right: Frank Oberle, Jr.

Winter

The real enemy is starvation

When the fields are covered with snow and the ponds are frozen, the moose turn to twigs and bark for nourishment. Insulated from the cold by a thick coat of hollow hair, the animals move to higher ground (above) or wander among protruding brush (right). A moose has no upper front teeth, so it must strip branches with its lips and lower teeth as this cow (below) is doing. By winter's end, the survivors are so run down that their sudden recovery in spring will seem almost miraculous.

Monkey Business

Facts and fancies about primates, from the smallest lemurs to the greatest apes

By Edward R. Ricciuti

MONKEYS and apes, lemurs and humans, galagos and lorises are all in the order, or biological grouping, called "primates"—a word meaning "those of first rank." This name tells us more about who does the naming than it does about primates: in the evolution of mammals, primates are not the first or final achievement, but rather an intermediate step. In many ways primates are more primitive than rodents or hooved mammals.

Primates fascinate us; we watch them with the special interest accorded relatives. They are a diverse group of about 200 species, ranging in size from the 1½-ounce mouse lemur to the 600-pound gorilla. Primates are divided into two suborders: the prosimians, which include seven families of small simians such as lorises and galagos; and the more advanced "higher" primates, consisting of five families of monkeys, the lesser apes (gibbons), great apes (chimpanzees, gorillas and orangutans) and humans.

The prosimians are the more primitive group. They are mostly nocturnal, secretive, and as a rule not as intelligent nor as facile as monkeys. Monkeys tend to be larger than prosimians, with well-developed brains and senses, and with hands capable of complicated manipulation; some New World monkeys even have prehensile tails as a sort of extra limb. The apes are in general the largest primates. All are tailless, have longer arms than legs, and except for orangs, can stand upright with ease. They are said to be the smartest of the primates.

The next few pages present a selection of assorted facts about the sometimes familiar, sometimes bizarre order we belong to. In some of these tales we may see reflections of ourselves. In others, we may see differences that could provide even better clues to that mystery called human nature.

The world of primates ranges from the familiar to the bizarre. The baboon (above) and the uakari (right) are both relatives of man.

148

Monkeys

Staying Single

Orangutans are strange and solitary animals, living reclusive lives in the forests of Borneo and Sumatra. Their name derives from the Malayan words *"orang utan,"* meaning "forest man." Early Malayans believed that orangs possessed the intelligence of people, and could speak, but did not because they were afraid of being put to work.

Adult orangs are loners, meeting only to mate. Females usually bear one youngster and raise the slow-maturing infant for six or seven years. The young are weaned when they are about three or four years old and for another year or two will remain with Mom, but gradually the orang will move away from home and establish its own territory, becoming yet another half-seen, half-legendary inhabitant of the jungle.

Listening In

The little aye-aye of Madagascar, one of the world's rarest animals, has a good ear for sound. The slender prosimian methodically taps trees and bamboo with its long fingers, listening for the change in resonance that indicates a beetle grub's hollow below the surface. Once the aye-aye locates a burrow, it chisels through the wood with its squirrellike teeth, then extracts the meal from the hole with its long, clawed third finger.

But Is It a Boy or a Girl?

The primitive aye-aye has long been in danger of extinction. In 1965 scientists transplanted nine of the lemurs from the island nation of Madagascar to a tiny islet called Nosy Mangabé to keep them in a protected habitat. In 1983 there was the first indication that the experiment was a success—scientists spotted a baby aye-aye in the island's thick forest.

Spider Monkey

Orangutans lead solitary lives in the forests of Borneo and Sumatra. The mother and young (above) will go their own ways when the youngster matures.

The cotton-topped tamarin (above) is a fierce hunter of birds and rodents.

Aye-ayes (above), primitive prosimians, are close to extinction in Madagascar.

Southern Asia's tarsier (above) is well adapted to its nocturnal life: its huge eyes spot prey easily in the dark.

Malayans believed that orangs were intelligent and could speak, but did not because they were afraid of being put to work

Always Look Back

The big-eyed tarsier, a simple primate of southern Asia, can swivel its head almost 360 degrees, making it one hard critter to sneak up on.

Watch Out, Batman

The owl monkey, or douroucouli, of South America is the world's only nocturnal monkey. Its immense eyes are so well adapted to seeing in the dark that it often catches and eats bats.

Skin Diver

The Philippine macaque, also called the crab-eating monkey, ranges throughout southeast Asia, inhabiting coastal swamps, forests and beaches. It is an excellent swimmer, able to dive into the water and catch crabs on the bottom.

Pint-sized Predators

Marmosets and tamarins are fierce predators, even though the giant among them is no larger than a gray squirrel. These small South American monkeys are said to catch and kill birds and rodents by grabbing the victim, often around the neck, and sinking their large canines into its skull. The *coup de grace* is not instinctive: older monkeys teach it to youngsters.

The Big Schnoz

The proboscis monkey of Borneo gets its name from the oversized nose that males grow as they mature. Most of the time this great schnozzola droops over the monkey's mouth like a deflated football, an appendage so large that the

he crab-eating monkey (above) is a primate with a taste for seafood. Living ong swamps and beaches in Southeast Asia, it is an excellent swimmer.

Colobus Monkey

151

Monkeys

creature has to shove it aside when eating. The nose has a special purpose, however: it amplifies the male's alarm call, a resonant "honk." When the creature calls, the nose balloons up in a remarkable simian imitation of Jimmy Durante.

Sky Diver

When Africa's black-and-white colobus monkey wants to get to a lower branch in a hurry, it performs a primate sky-diving act. Leaping into the air, the monkey spreads out its arms and legs so that its lush, long fur fans out. The fur probably acts like a parachute, braking the monkey's fall, though the animal may drop as much as 50 feet.

Swingers

Gibbons have elongated fingers that readily hook over branches in the jungle canopy. When a gibbon wants to get someplace in a hurry, it swings with lightning speed from branch to branch, grabbing and releasing its hold so quickly that it spends more time in the air than gripping the branches. In this manner gibbons can throw themselves from tree to tree so easily they appear to fly. Gibbons also use branches as springboards, bouncing off them to hurtle through the air between widely spaced trees.

At the Watering Hole

Gibbons spend a lot of time hanging around. They dislike coming down from trees and even drink while hanging from a branch or vine over the water. A thirsty gibbon in this position dips its long arm into a stream and then drinks the collected moisture from the back of its hand.

Campbell's Monkey

ROD WILLIAMS (BRUCE COLEMAN INC.)

Only the male proboscis monkey grows a gigantic nose (above), because only the male emits the honking alarm calls which are amplified by the appendage.

ibbons are noted for their reach; tip
tip, the silvery gibbon's arms (above)
re longer than the gibbon is tall.

The fleshy nose of Borneo's proboscis monkey is so large that the creature has to shove it aside in order to eat

Guenon

Monkey Tails

Despite the common image of them, few monkeys can swing by their tails. Those that can all live in the New World, and of these, the spider monkey has the tail most like an extra hand. It can use this appendage to pick up objects, such as fruit, and a spider monkey owned as a pet by a scientist even learned to open a door by the handle with its tail. When a spider monkey happens to fall from a branch, it almost never hits the ground. If its hands or feet do not grab onto a limb before it drops very far, its tail does, serving as a convenient sky hook.

Fire in Their Eyes

Lemur was the Roman name for the spirits of the dead who roamed the night, their eyes burning balefully and voices moaning. The lemurs of Madagascar, an island off southeastern Africa, were named by French colonizers. Many of the monkeylike prosimians are nocturnal, with a call like the laments of lost souls. Their eyes seem to glow in the dark, due to a lining called *tapetum lucidum,* at the back of the eye.

Biggest of the lemurs, about a yard from the nose to rump, is the indri. Its ancestors were even bigger. Indri skulls the size of a donkey's head have been found in Madagascar, and scientists believe that man-sized indris inhabited the island at the time of the first human settlers.

The indri got its name in the seventeenth century, when a French explorer saw natives point to one of the creatures in the trees and shout, "Indri!" It was not until later that the word was translated: the native was saying, "There it is!"

Reach

The siamang, biggest of the gibbons, is about three feet long from head to rump; its immense arms can spread out past five feet.

nlike most primates, the ring-tailed lemur of Madagascar (above) lives on
e ground. It is found in the treeless mountain slopes of the island.

153

Monkeys

Whiskers

The lion-tailed macaque, a native of dense mountain forests in southern India, gets its name from the leonine tuft of hair at the end of its long tail. The resemblance does not end there, however: the macaque's face is framed by a great mane of long, whiskery hair.

All in the Family

Unlike most primates, gibbons are often monogamous. Father, mother, and young live in family groups with offspring of various ages.

The Other Chimp

Most people are familiar with the common chimpanzee, star of circuses and movies. What they don't know is that the common chimp has a smaller, fascinating cousin, the pygmy chimp, or bonobo. The pygmy chimp, native to the remote central jungles of Africa, was not discovered by scientists until 1929. This chimp has rounder eyes and a less protruding jaw than the common kind, giving it a more human expression. Pygmy chimps walk upright even more than their bigger cousins, and seem to be gentle and intelligent, using a wide variety of gestures to communicate. For these reasons and others, some scientists theorize that the pygmy chimpanzee is man's closest living relative.

Big Baboon

The male mandrill baboon of West African forests has a scarlet snout and blue and purple cheeks (see front cover). Despite its clownish appearance, however, it is not to be trifled with. A big mandrill weighs more than 100 pounds and has the fangs to match. Mandrills have been known to stand up to leopards.

Who Gets the Best Seat

Gorillas live in stratified social groups, in which silver-backed males rank above other males, all adult males rank above females, and females above juveniles. One scientist, observing a group in the rain, saw a young gorilla take shelter under a tree, only to be pushed

The threatening expression of the lion-tailed macaque (above) is well framed by its leonine mane of hair.

Tradition has it that the British will hold the Rock of Gibraltar only as long as the Barbary apes are prospering

Barbary apes (above) are traditional inhabitants of the Rock of Gibraltar.

Mandrill

Gorillas (above) live in stratified groups, in which a silver-backed male (rear of picture) is always dominant.

These vervets (above) are also called green monkeys for their tinted fur.

out by a larger female. No sooner had she sat down, however, than she, too, was pushed back out into the rain by an adult male.

Instant Umbrellas

Gorillas aren't the only apes to dislike getting wet; when it rains, orangutans often cover themselves with large, broad leaves to keep dry.

Camouflage

When frightened, the little vervet monkey of Africa heads for the trees. If these do not provide enough cover, the monkey grabs leafy branches and pulls them together around itself.

Tradition

The Barbary apes—really macaque monkeys—of the British colony of Gibraltar have been on the Rock for at least 1,300 years, and probably much longer. Although there is a slim chance they arrived at Gibraltar naturally from North Africa, they were most likely introduced there, perhaps by the Romans.

However they got there, tradition says that if the apes ever leave the tiny enclave in the south of Spain, the British will follow. Only as long as the apes remain will the British retain their hold on the strategic Rock. Therefore, whenever disease or other pressures have diminished the ape population, the British have captured more of the beasts in North Africa and freed them on Gibraltar. ∎

Edward R. Ricciuti is a former editor of the magazine Animal Kingdom.

Douc Langur

155

DESERT

MACHINES

For centuries camels have given people milk, shelter, rope, fuel, meat, leather and transportation... they still do ♩

Ungainly but useful, the one-humped camel, or dromedary, has changed the lives of people in most arid sections of the world. This nomad in Afghanistan is watering his animals, but camels can go up to three months without drinking.

"Neither snow, nor rain, nor heat, nor gloom of night stays these couriers from the swift completion of their appointed rounds."
—Quotation carved along the facade of New York City's main post office

BY RAY VICKER

THE author of those famous words was Herodotus, the Greek historian, writing in the fifth century B.C. He wasn't describing the United States Postal Service, of course, but the communications network of Xerxes, the ruler of Persia. Camels played an important role in that network, just as they have done in the social and economic lives of people in desert lands for centuries.

Now, as then, camels are life-giving desert machines. They provide food, shelter, transport, clothing and revenue. Nutritious camel milk, one to ten quarts per day from a lactating female, is enough to sustain human life. The hairs of the animal are braided into rope or woven into cloth for tents or robes. Its dung burns well when dried for fuel. Its skin provides a leather ideal for water bags, belts and straps. And some Arabs prefer its meat to beef.

Few animals have influenced people as much as this ungainly four-legged creature with the sad eyes, the frightening bellow, the long legs and neck, the vile temper. Certainly, the horse, the ass, the cow, the sheep, the pig, the goat, the water buffalo, the yak and the dog have all played roles in people's quests for an improvement in their life-styles. Yet few ever became as vital as the camel. Since its domestication some 3,000 to 5,000 years ago, it has performed tasks now handled by cargo ships and airliners. The nomadic Bedouins even called it "the ship of the desert."

The ancestors of this desert beast originated in North America, migrating via a land bridge at the Bering Sea into Asia. Two species survived, the one-hump Arabian, or dromedary, and the two-hump Bactrian. Another camel branch ambled south to South America, developing into the wild vicuna and guanaco, and the now-domesticated llama and alpaca.

The dromedary has found its niche in hot, desert lands in Africa, the Middle East, Pakistan, Afghanistan, and northwestern India. The Bactrian's territory is Central Asia, Mongolia and northern China. The two-humped animal is shorter, more solid than the dromedary, has thicker body hair and is better adapted to cold. But the Bactrian camel has never been as important in human social history as the dromedary.

Contrary to the old myth, an Arabian camel "stores" no water even when it drinks 30 gallons at a time. It is no myth, though, that a camel may go three months without water during cool winter months. It gets all necessary moisture in plants that it eats. Moreover, its body excretes or sweats little water. A camel's body temperature fluctuates between 93 and more than 104 degrees F. to "regulate" and conserve body fluids. This adaption to dry country is coupled with great lifting strength, resulting in a near perfect animal for use ➥

In a desert "dairy" Rendille tribesmen in northern Kenya collect camel milk — high in vitamin C. Some tribes bleed their animals to produce an even more nutritious mixture of blood and milk. In Arab lands, Bedouins offer "milking privileges" to thirsty desert wayfarers.

Medford Taylor (2); pages 156-157; Victor Englebert

in the difficult conditions of the desert.

That is why Xerxes' couriers used dromedaries. In the arid wastelands of old Persia, royal retainers stabled camels at khans, or inns, in cities and oases along desert trails, ready for riders to remount with their leather mailbags after loping into courtyards on tired animals. The idea sounded so good to America's Congress in the 1850s that the United States almost adopted a camel rather than a pony express to connect California with the rest of the country.

By Xerxes' time, camels had already established their place in Old World history as beasts of burden, mounts, sources of food and providers of fiber. Even today, nomads, whose ancestors once moved caravans across Asia and North Africa, still rate wealth in camels. Some tribal laws still require bridal dowry payments and blood money in the animals. Blood money is the compensation a murderer pays to a victim's family to avoid blood feuds which otherwise might tear a tribe apart. The blood money might be a dozen or more camels for a murdered man in some tribes, with the value of a female victim set at half of whatever a man might be worth in like circumstances.

Not too many years ago, I learned just how much life in the desert depends upon the camel. I was riding across the desert in Saudi Arabia in a pickup truck with an official of the Arabian American Oil Company, when we encountered a lone Bedouin boy tending camels miles from any settlement. The December sun had lost its sting, but he apparently had no supplies with him, no water, no food. When I expressed surprise at his carelessness, the official braked and called to the boy. They exchanged a few words.

On his head, the boy wore a red checkered *keffiyeh* (termed a *ghutrah* in Saudi Arabia). Now, he whipped it off, revealing a tin bowl which sat on his uncombed hair like a skullcap. "He milks the camels when he is hungry or thirsty," explained the official. "He could live out here for days if necessary."

Camel milk is high in vitamin C and is believed to have curative powers. In Central Asia, it is prescribed for stom-

Fuel in the Sahara is hard to come by, and a Tuareg girl in Niger collects dried camel dung. It will be burned for cooking. Virtually everything about the animal is useful. Skin and hair, for instance, are made into products ranging from water bags to cloth for tents.

ach ailments and for tuberculosis. In Africa, tribes like the Galla of Kenya bleed their camels to provide a blood-milk combination which is even more nutritious. In Arab lands, Bedouin hospitality offers milking privileges to any wayfarer caught in the desert without nourishment when camels are around.

Camels have other uses, too. In the Middle East, camel dung is mixed with herbs and employed for dressing wounds. *The Medicine*, an eleventh-century Arabic medical tome, claimed that the fat of the camel's hump is good for dysentery, its bone marrow cures diphtheria and its dried brain is a treatment for epilepsy.

Arabs have a thousand words in their vocabulary to describe this multitalented creature. Many of them are endearing enough to apply to a sweetheart. The camel's cheeks are said to be as

South America's humpless camels

THERE is almost a mystical affinity between the Andean Indians and the camellike animals of their high-mountain lands. Four species share a common North American ancestry with the Arabian and Bactrian camels, and they have provided the Andean Indians with transport, meat and wool since time immemorial.

These four humpless "camels" of the Andes are the guanaco, vicuna, llama and alpaca. The domesticated, cargo-hauling llama has a long, graceful neck, erect ears and large, doelike eyes. An adult weighs from 250 to 400 pounds, stands 3½ to 4 feet high at the shoulder and carries from 60 to 100 pounds of baggage for about 15 to 25 miles a day. The alpaca, which is also domesticated, is smaller than the llama, but its body has so much wool that it looks like a walking pile of fluff. Both the guanaco and the vicuna are wild. The guanaco is also smaller than the llama, with a more graceful body. The vicuna is the smallest of the four, with the

Guanaco

Alpaca

Vicuna

Llama

trim lines and speed of a gazelle and silky wool four times thinner than human hair, yet with thermal qualities better than the best sheep's wool.

Today in Peru, vicunas are making a comeback after a brush with extinction, and in the United States, strangely enough, the llama is winning a toehold. There are about 3,000 llamas in the United States, descendants of animals imported before trade in the species was banned in the 1930s. Western resort and pack trip operators find the animals ideal for lugging loads formerly carried by mules. Meanwhile, in South America, breeders have attempted to cross alpacas and vicunas. Their creation is called the paco-vicuna, an animal that gives nearly as much wool as the alpaca, but with a quality not much inferior to that of the vicuna. So far, however, developers of this strange conglomeration of South American "camels" have struggled to make it breed true. — *Ray Vicker*

Time off for caravan camels in Niger comes after a grueling workday. These "ships of the desert" munch on forage, although they also relish desert plants. Camels can pack loads of 600 pounds for up to 30 miles a day, but the freight caravan of old has all but disappeared.

soft as silk, its eyes like those of a young maiden and its neck like a slender minaret. Of course, Arab camel drivers have their swear words, too, often uttered with loud shouts which start camels bellowing like frightened cows.

Perhaps the most illogical joke ever concocted involves the camel: "It is an animal created by a committee." But the camel is anything but a biological misfit. It is more like a slow truck which needs no gasoline, lasts for up to 40 years and rides across roadless terrain. It can smell water six miles away when the wind is right, locating water holes missed by people. It needs no fodder, but relishes cacti, thorns and salty bushes. It can haul a half-ton of freight for short distances, more than an elephant manages. For more normal distances — 20 to 30 miles a day — it pulls loads of 600 pounds.

No wonder the camel became the ship of the desert. Drivers joined long strings of camels into caravans of several thousand animals, their bells tinkling across the desert. Previously, trade moved only on flimsy ships, and then not too safely. The camel could go almost anywhere in lands then known. So trade flourished, spreading products and ideas, and energizing civilizations.

Caravans transported ivory and skins from Africa, silks from India, paper from China, spices from Southeast Asia, and frankincense and myrrh from southern Arabia. The aromatics are from gum resin scraped from tree trunks in what is now Oman and South Yemen. Greeks, Romans and others burned tons every year in their religious rites, and used them in the manufacture of perfumes and in embalming dead. The Queen of Sheba may have traveled via camel over this Incense Road when she made her famous visit to King Solomon, for her kingdom occupied the green, terraced highlands of what is now North Yemen.

The Incense Road already was long established when the Nabataeans gained control of it in the fourth century B.C. From their capital at Petra, "the rose red city half as old as time," they dominated a traffic equivalent to a quarter

For desert travel, the traditional camel is still tops, even though its riders sometimes succumb to other innovations — such as sunglasses. Like these Tuareg nomads, though, all desert people are facing change, and that has meant a changing role for dromedaries, too.

of the annual gross national product of the Roman Empire. Their caravans traversed most of the Middle East, with each venture operated as a stock company owned by merchants sitting in the capital.

All that camel traffic speeded decay of the network of 50,000 first-class and 250,000 miles of feeder roads across much of the Roman Empire, but not by overuse. After Rome fell camel caravans proved so effective without roads in the Middle East and North Africa that nobody bothered to maintain the Roman system.

The *Bible*, in one of its 31 references to camels, says that Job of the Old Testament owned 6,000 of the animals. Abraham, too, had a string (or "mob") of them. But the camel has always been more closely associated with Islam. Mohammed, the prophet of Islam, was a camel driver and trader who is said to have made at least one caravan trip from Mecca to and from Syria.

When Mohammed appeared on the scene, a century after Rome's fall, his people were ready for conquest of the southern half of the known world. Camel corps aided Moslem warriors in their sweep to power. In a crucial battle with Byzantines at Yarmuk, Islam triumphed when a blinding sandstorm covered the battlefield as if given by Allah for the camel riders of Arabia. No animal can maneuver as well in a sandstorm as a camel.

While one Moslem army swept westward as far as Tours, France, another penetrated Central Asia, defeating a Chinese force on the Talas River in 751 A.D. Moslems captured numerous prisoners, some Bactrian camels and loot which included a strange substance — paper. Chinese had long known how to make it, but the Middle East and Europe had to be satisfied with scarce papyrus and parchment for writing.

Some of the Chinese captives knew the secret of papermaking. They gave this knowledge to the Arabs, who began making paper in Samarkand for shipment to Damascus. That city then developed as the papermaking capital of the world, with tons of the stuff

moving annually via camel train to all parts of the expanding Islamic Empire. One frayed letter written in the eleventh century reports that 28 camel loads of Damascus paper were being shipped from there to Egypt as a single order.

For centuries, the only way to travel through much of Asia or North Africa was via camel caravan. When Marco Polo started across Asia from Armenia in 1272 for the court of the grand Khan in China, he probably traveled by caravan. Napoleon was so intrigued by the camel when he conquered Egypt in 1798 that he formed a French camel corps and apparently considered introducing the animal to Europe. He returned to France with a riding camel of his own, but presented it to a zoo.

The French, however, didn't outgrow

A camel is like a slow truck which needs no gasoline, lasts for up to 40 years and travels almost anywhere

---

their love of the camel. French camel corps helped establish many of that country's colonies in Africa, including Algeria in 1835. Britain created a camel corps in Sudan which operated there until 1921. Musty records in the British Museum indicate that during the Crimean War in 1853-1856, both the Russian and the British armies had trains of camels. One report notes that "25,000 camels were serving in baggage trains of General Napier."

The camel has a place in American history, too. It was in 1836 that somebody in Washington conceived the idea of utilizing the camel in the West. Nothing came of it until the Mexican War added 529,189 square miles of new territory to the Republic. Then, the camel lobby gained strength, with support for the Camel Express coming from California. In 1855, Congress appropriated $30,000 for purchase of 75 camels from Mediterranean lands as an experiment in their usefulness in the Southwest. The War Department under Secretary Jefferson Davis took charge of the program.

In two voyages, 75 Arabian camels

were landed at Indianola, Texas, and marched to an encampment near San Antonio designed by Colonel Robert E. Lee. There a contingent of them was consigned to Lieutenant E. F. Beale for surveying a wagon road from Fort Defiance, New Mexico, to Southern California. As the project progressed, Beale became a strong supporter of the camel, declaring that it could do three times the work of an army mule without complaint. But the mule skinners of the American Army didn't like the imports, and camels frightened every horse they met. Several camel drivers who had been brought to America deserted, and the camel experiment petered out when the Civil War exploded.

In 1864, the Army auctioned its camels, most going to mining camp draymen. For years, the camels worked in Nevada mines, hauling ore to mills and frightening horses. In 1879, the Nevada legislature voted to bar camels from transport in the state. The animals were allowed to run wild and most were subsequently shot in Arizona or Nevada.

It was an ignoble finish for a glorious experiment. But the camel really couldn't compete with railroads. Nor can it usually compete with trucks today. Now, even in the Middle East and North Africa, where camels are as common as ever, only the poorest nomads still use camels for transport — tribes like the Rendille in Kenya, the Somali in Somalia and various Sudanese groups. Today, the animal often does donkeywork, surrenders its milk, then has the slaughterhouse for its finish.

But it is still revered, and if the great camel trains are all but dead, the camel, at least, remains an imposing — and useful — presence in some parts of the world. Near Egypt's pyramids, camels still trudge endlessly in a circle at waterwheels, lifting water from ditches to irrigate small gardens. On a Delhi street in India, camel-drawn carts clog traffic, some bearing ten-foot-high loads. And in Amman, Jordan, near the Basman Palace, a troop of red-kerchiefed Desert Police still parades before King Hussein on well-trained camels. ∎

As a senior foreign correspondent of the Wall Street Journal, *Ray Vicker traveled widely in the Middle East where he often observed people's relationships with camels firsthand — sometimes from a camel's back.*

PHOTOGRAPHS BY STEPHEN J. KRASEMANN

Hail, hail, the gang's all here

THEY arrive in a timid mass, swimming back and forth along the empty shoreline, too wary to approach the beach. Finally, one brave bull makes a move. Grunting and groaning, he heaves his 3,000-pound hulk out of the sea and settles atop a barnacle-encrusted rock. Almost immediately, other bulls waddle ashore behind him. Soon, the entire beachfront on Round Island is a throng of Pacific walruses, huddling together and sunning but ready to retreat into the water at the slightest provocation.

Each summer, some 15,000 bulls gather on this tiny speck in Alaska's Bristol Bay for a rollicking bachelor party. Only when winter comes do these mighty behemoths migrate back out to distant ice floes, where they join the cows and calves that have summered farther north. Round Island is the only place on the American side of the Bering Sea where large numbers of male walruses regularly "haul out," as their landing is called, each year; many more of the animals summer on the Russian side of the Bering.

Just three decades ago, the Pacific walrus population was only about 50,000. Today, some scientists believe it may be at least 200,000, or as many of the animals as there were a century ago. "They are hauling out in areas where they haven't been seen in a long time," says Francis Fay, a University of Alaska biologist. One reason for this remarkable turnabout is the Marine Mammal Protection Act of 1972. That landmark law restricted walrus hunting to the subsistence needs of Alaskan Eskimos, Indians and Aleuts. It also transferred the responsibility for taking care of the creatures from the state of Alaska to the federal government — a provision that set off a furor.

Arguing that it was better equipped to manage the animals, the state regained its authority in 1976. It then imposed a ceiling on the number of walruses that subsistence hunters could kill. But now the controversy has come full circle, for last summer the state gave up its hard-won control over the walrus. Reason: a federal court ruled that neither the state nor the U.S. can legally limit native subsistence hunting — a power the state feels it must have to manage the animals. As a result, the U.S. Fish and Wildlife Service has reluctantly resumed its on-scene monitoring of the natives' walrus hunt. The aim is not to limit the kill but to make sure that the animals are not, as the law puts it, "taken in a wasteful manner."

As for the bulls of Round Island, they are blessedly unaware of who controls their fate. Right now, in fact, they are about to undertake their journey back out to sea. This year's bachelor party is just about over. — *Mark Wexler*

Like duelers "presenting arms," two Round Island bulls thrust out their tusks in a test of dominance. Often, the animal with the greatest bulk and tusk size wins by intimidation. When a fight does occur, only occasionally is one of the combatants seriously injured. Walruses are protected by two-inch-thick skin and skulls so dense that the animals use them to smash through ice a half-foot thick.

One of the world's longest-running beach parties takes place each summer and fall on this narrow Round Island peninsula, where walruses cover every foot of space. The animals' survival often depends on "bundling," or huddling very close together, to share body heat. Even during the summer months, they are highly social.

Though the walrus spends more time on land than most other marine mammals, walking is an ordeal for the 3,000-pound giant. Pivoting its hind flippers, it advances slowly in a series of lurches.

A simple task like scratching can be a major chore for a bull when out of the water (top). In the ocean depths, however, the animal is surprisingly agile. The walrus feeds primarily on clams and other shellfish, and can remain submerged up to ten minutes at a time. Just before it surfaces, the creature exhales (above), gathers in another breath and then dives again. At sea, walruses sleep by swelling their neck pouches with air and floating like bobbers (opposite).

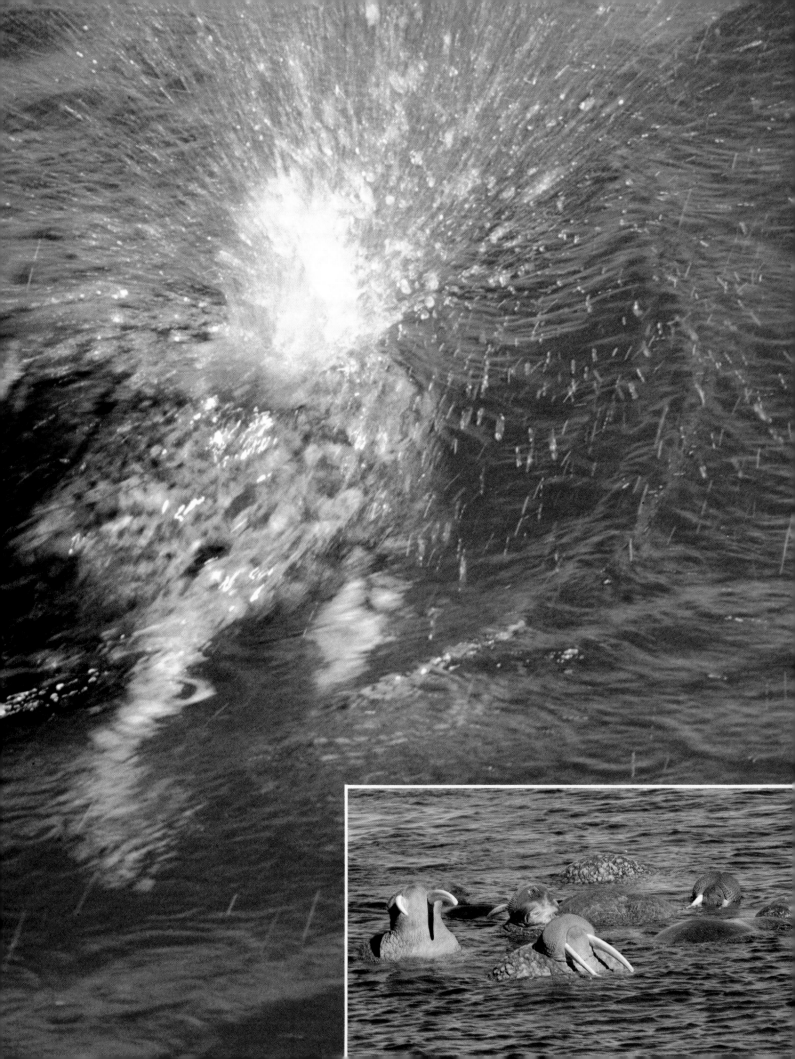

Leonard Lee Rue III

There's nothing gentle about a buck in rut

TAKING OFF THE

BY GEORGE H. HARRISON

L AST fall, a young Ohio Division of Wildlife photographer, Kim Heller, was gored by a buck deer while he was photographing the animal at a private game farm. Three days later he died. During that same week, I was photographing a buck deer in a research pen at Pennsylvania State University. I also was attacked, by a 350-pound whitetail, affectionately called "Sam." I escaped only because of quick action by a researcher, Rick Potts, a graduate student who had worked with Sam for several years.

We had entered the two-acre wood-land pen looking for Sam and were caught off-guard when we found him behind us, blocking an escape to the gate. It was immediately obvious that Sam was going to attack. The enormous buck trotted to within ten feet, snorted, pawed the ground, lowered his head and huge set of antlers, and charged us like a shot out of a cannon. Rick was ready, but even a blow across Sam's nose didn't stop the animal from knocking the biologist to the ground. I stood in disbelief behind Rick with nothing but my camera to protect me. Luckily, Sam's antler spread was so wide that Rick fit between the points and was actually butted by Sam's forehead. One antler point did scrape Rick's leg. Quickly, Rick got back on his feet and distracted Sam with quiet talk long enough for us both to escape.

It is not unusual for buck deer to turn mean enough to attack humans this way. It's particularly true of "tame" deer kept in captivity for research or as pets by people who forget that deer are wild animals. Each fall, when the bucks come into breeding condition, they change dramatically in both body and personality. A precursor of this Jekyll-to-Hyde transformation comes in late summer, when the bucks rub the soft mosslike covering off their antlers.

The antlers themselves are not horns.

174

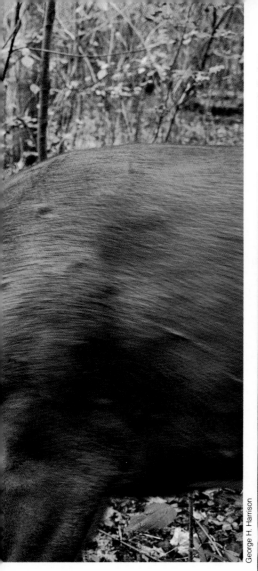

George H. Harrison

It begins with an itch to get rid of the mosslike covering, or "velvet," that sheaths and nourishes a male white-tailed deer's annual set of antlers (far left). When the rack stops growing, the membrane dries and a buck rubs it against brush and saplings (center). Long after the last velvet is gone, the animal keeps polishing. Then, as the November rutting season nears, this activity turns into mock combat. The rut itself is at first a time of frustration for the males — the does won't begin mating until December. Meanwhile, its antlers finely polished (below), a buck is very edgy and dangerous — sometimes even to humans.

VELVET

They are solid bone structures made up of calcium and chemical salts which are shed each year following the breeding season. Antlers are also grown by moose, elk and caribou — all members of the deer family. Until recently, researchers didn't know what biological processes were involved in growing the bony headgear. Experiments at Pennsylvania State, in the same research pens in which I was attacked, have shown that the pineal gland in the deer's brain serves as an "internal clock." The gland tells the buck when to grow antlers and when to change into his winter coat. The antlers of well-nourished bucks grow at a phenomenal rate — perhaps as much as half an inch per day. Their size and number of points have more to do with diet than with age. This "rack," as a set of antlers if often called, seems to be as much a status symbol as a means of fighting over does. (In encounters with predators, the hoofed forefeet are a deer's primary weapon.) The growing antlers are soft, nourished by a covering called velvet. By late summer, they stop growing and harden.

That's where the "rub" comes in. After the growths have stiffened, the bucks spend a month or more polishing them on the trunks of small saplings and brush. The scrapes they create on trees are called buck rubs. Most of the job is usually completed within a day or so, but shreds of velvet may cling to the antlers for several weeks. Some deer eat the covering as it sloughs off. By November, the velvet is gone and the buck reaches breeding condition; it is now, as naturalists say, in the rut.

After the antlers are polished, the buck is a changed beast. Even the tamest individual becomes a wild and dangerous animal, especially if he is in a pen and has no does to satisfy his mating urge. To trust him is to flirt with disaster. ■

George H. Harrison is a field editor of this magazine.

175

A school for elephants

BY DENIS D. GRAY

PHOTOGRAPHS BY NEAL ULEVICH

SOOOONG!" shouts the chief of mahouts. As quickly and nimbly as pointer dogs, a dozen elephants lift their right forelegs and 12 mahouts swing up to lodge themselves behind huge flapping ears and powerful necks. Two younger animals, a good 800 pounds apiece, crack each other playfully on the head, oblivious to the sounds of crisp, morning discipline. An older elephant scoops up a bouquet of branches and deftly swats at the flies buzzing around its skin.

It is two hours after sunrise at the Young Elephants Training Center in northern Thailand — two hours after the students have lumbered out of the hilly jungle surrounding the school to be scrubbed down in a mountain pool and preened for the morning's training. Today, like most days, these elephants will practice pulling teak logs with chains strapped to their massive bodies. They will lift and stack logs weighing a ton or so into neat piles with their tusks, and they will work through an impressive repertoire of some 26 commands. "The oldest student here is ten," says Nuon Komrue, the chief of mahouts, riders and keepers of elephants in Asia for generations. "He will graduate soon and go to work in the teak forests."

Meanwhile, 330 miles to the south in the bustling Thai capital of Bangkok, a trainer holds up a stick of sugar cane to prompt Plai Boon Rod to lift his trunk in royal salute. The rare, three-year-old white elephant, a symbol of royalty, is to be presented to the King of Thailand amidst the chanting of Buddhist monks and the anointment of its frizzy head by his majesty King Bhumipol Adulyadej. Born in the jungle and discovered to possess such albino characteristics as whitish

eyelashes and toenails, Plai Boon Rod has been given the exalted royal title of *Phra Sawet Suthavilas Attakhachartpitsanupongse*. He will live out his life in honor, pampered and featured at court rituals in a tradition that goes back to the days when Burma and Siam fought a war over a white elephant and lullabies were sung to the beasts.

Elsewhere, a Bangkok sales executive points to a yellow, 20-foot-long piece of machinery with a stubby blade up front, a powerful winch in the back and four massive, deep-tread tires. The tractorlike, American-made vehicle is called a "skidder." The salesman ticks off its advantages: a medium-sized skidder can pull logs weighing up to 20 tons; a full-grown elephant can manage no more than two tons. Like an elephant, the machine can climb precarious slopes and swivel around trees. It takes three to four days to train a skidder driver while elephant schooling lasts about seven years. And the skidder, moving faster than an elephant, never tires. True, a medium model skidder costs about $75,000 while a fully trained elephant in Thailand currently fetches some $4,000, but in the long run the machine pulls ahead in cost effectiveness. The executive says 250 units of his company's model have been sold in Thailand and the demand rises year by year.

Only a few decades ago, local foresters and the European teak dealers would invariably have concluded that elephants were "the most lovable and sagacious of all beasts." Able to move up slopes of up to 70 degrees, push through thick jungle and break up log jams in rivers and streams with powerful head and trunk thrusts, the elephants made possible a multi-million

dollar industry in remote stretches of Asia during the last century and the first half of the 20th. Consequently, they were pampered, doctored and normally loved by the men who often spent the bulk of their lives with them in the forests of India, Burma, Thailand and elsewhere. Now, the Thai skidder salesman predicts that it is only a matter of time before machinery will totally replace elephants as logging workers in Thailand, and eventually in all of Asia.

Is the age of the elephant in Asia really passing? The statistics seem to say so. In Thailand, the domestic elephant population has dropped, according to government statistics, by some 5,000 since the end of World War II. And in Burma, which has nationalized all pachyderms, there are now some 2,700 working elephants, compared with about 5,200 employed by British and other foreign firms before the war. Even Burma, with an economy in shambles, has purchased some skidders. The denuding of forests and continued poaching for ivory and sometimes meat in many corners of the Asian elephant's habitat have also reduced the size of wild herds, from which domestic stocks are replenished.

The Asian elephant — lighter and smaller than its African cousin — still runs wild in India, Burma, Thailand, Vietnam, Laos, Cambodia, Indonesia, Sri Lanka, Malaysia, Nepal, China, ⤵

Preparing for a "career" in forestry, young elephants at a training center in northern Thailand learn to lift logs with their tusks and trunks. But the burly trainees may find their jobs usurped by tractorlike skidders.

Overleaf: Mahouts, as the elephant trainers are called, form bonds with their animals and may move on with an elephant following graduation. Training lasts for seven years.

There are no dropouts
or flunkouts and the graduates
work for peanuts

Stacking logs in neat piles is a difficult but basic skill young elephants must learn.

Relaxation is part of the school regimen, along with summer vacations and days off.

Cooling off at a nearby stream, pupils get a rubdown from their mahouts. Most mahouts ascribe "spiritual power" to the elephant, which is revered generally in Thailand and is entwined in the country's culture.

Teaching aids include parallel rails through which an elephant must push a log with its trunk. Shouted commands are accompanied by gentle kicks, rubs or knee thrusts. The animal learns basic orders within a month.

Bangladesh and possibly other areas. Last year, an Asian elephant study group estimated the wild population to be between 34,000 and 45,000. But at a meeting called to discuss the situation, trouble was predicted. One conference paper even warned that wild elephants in Thailand would be seriously threatened with extinction in the next 10 years.

But at least one member of the study group, Jeffrey A. McNeely, an American conservationist who has worked in Thailand for almost a decade, believes that the elephant, as a working animal, still has a future in an energy-short world. "Elephants run on local fuel, reproduce themselves and still can go into some places that machines can't," he says. "Elephants are going to be around longer than petroleum. They've also shown an amazing ability to adapt to man in Asia."

Certainly the training school for young elephants in Thailand seems unfazed by dire predictions. Here, the centuries-old bond between man and pachyderm still exists. With less than a hundred pounds on his frame, Nuon Komrue is dwarfed by the animals around him. But they are his life. The chief mahout says he has worked daily with them for 22 of his 49 years, ever since he traded his pans and ladles for a bull hook. The son of a poor rice farmer, Nuon started as a cook for the Forest Industry Organization, the government department which runs the school. Taking a liking to the animals, he talked his way into a job as a *Khon Teen* — a foot-person charged with performing some of the more menial tasks of elephant care without actually being allowed to ride the animals. He became a mahout two years later.

Although some groups in Thailand — especially a few tribes living along the Cambodian border — have been elephant men (there are no female mahouts) for generations, those at the school and others in northern Thailand come from varied backgrounds and don't necessarily pass on their skill to their offspring. Nuon says many of his fellow mahouts are poorly educated but that because of the training with elephants, their status rises above that of a laborer, a taxi driver or a sharecropper.

Slight, with whiskers and a wispy beard, Nuon smiles broadly as he describes his life. He rises daily at five except on Buddhist holidays — days of the full and half-moon — when the school is closed. He eats a bowl of sticky rice, sometimes graced with bamboo shoots or other vegetables from his garden. Then he strides into the forest with the other mahouts to locate the school's students, who are let out to feed and rest each night.

It's about an hour's walk, but Nuon says it isn't any trouble finding the herd. The men, he says, know their own animals so well that they can track them down by the characteristics of their droppings or the grass and leaves they chew and then spit out by the wayside.

"*Phe, phe,*" the mahouts call out to the elephants. "Come, come." Dutifully, the elephants saunter out of the thick green to their masters, ready to be mounted and ridden out of the jungle to the school.

The school is the only one of its kind in Thailand and, for its eight years of existence, it has built an enviable academic record as far as most educational institutions go. It has never had a dropout and its director, Dr. Chaum Kinvudhi, says he can't remember teaching a stupid student.

While private entrepreneurs, villagers and some tribes in Thailand still catch wild elephants and train them,

With heavy chains strapped to their bodies, pairs of elephants learn to pull teak logs. Teamwork is the most difficult part of the training, says the chief mahout, although the elephants seem to come by it naturally.

Banana leaves and stalks are part of the diet for baby elephants although mahouts may give them special treats such as sugar cane. Older elephants consume 300 pounds of vegetation and 50 gallons of water daily.

without benefit of "formal schooling," the school's recruits come from the 100 or so animals that work for the Forest Industry Organization, which operates various logging sites around the country. A pregnant mother from one of these sites is put on a ten-wheeled, specially-cushioned truck and ferried to the school in time to give birth to a baby that automatically becomes enrolled. The baby stays with its mother at the school for three years or so and then undergoes what the mahouts describe as a painful separation. The mother goes back to work in the forest, and the child begins training which will last for the next seven years.

The young elephant is placed in a tight pen made of logs and with the help of sugar cane and other elephant delicacies is taught the basic commands. It also becomes familiar with hobbles and dragging gear, and with having a mahout perched on top. The school director says the elephants can pick up the basic commands in two weeks to a month.

Out of the pen, the next year is spent in the basics of logging work and a daily routine on the training ground: walking in file and in pairs, stopping, getting down on all fours, lifting and pulling. The commands are shouted out by the mahouts in a combination of Thai, a local Thai dialect and the language of the Karen, the great elephant keepers of Burma and Thailand. Each verbal command also has its equivalent in a subtle, if not gentle-seeming, kick, rub or knee thrust by the mahout in the region of the elephant's ears and neck. On command, a column of elephants will wheel around in sharp precision, and a bull elephant will thrust his tusks under a log and start lifting without a sound being uttered.

Teamwork, the director says, is the most difficult part of the training but elephants seem to come by it naturally. In the wild, elephants have been seen to sandwich a baby between them as protection against tigers or, working as a pair, to ferry a young one across a swift stream by using their trunks as rafts. When a mother gives birth, a companion, called an "auntie," is on hand to act as midwife, tearing off the membrane in

Baby elephants leave their mothers at the age of three to begin school. Their mahouts may be poorly educated themselves, but they acquire elevated social status. Both mahout and elephant will retire at age 60.

182

which the 200 or so pounds of baby comes, and helping the newborn in breast-feeding.

The teamwork at the school is designed to get as much elephant muscle behind a certain task as possible. A heavy log may be chained to two or more elephants who heave in unison. Or a trio of animals may move up on a log, scoop it up simultaneously with their tusks and lay it gently on a pile, taking care that no member of the team gets ahead or behind another member.

After passing through primary and secondary training at the school, the elephant, aged about 10 or 11, is ready for work on one of the government's forestry projects, hauling sawed logs from timber sites through

the jungle to yards where the logs can be transported by trucks, train or water to the markets. This is the battleground between elephants and evermore sophisticated logging machines.

Both during training and in work, the "elephants union" is strong. The government's animals train and work only a few hours a day and then are let out to roam and gorge themselves with their daily intake of some 300 pounds of vegetation and as much as 50 gallons of water.

There are weekly rest days, and from March until May each year there is "summer vacation camp" during which they do nothing but feed, sleep and mate. Retirement is mandatory, and medical care for the sensitive animals is prompt and pro-

fessional. A mahout will normally take the same elephant through training and move with it after graduation. The pair may stay together for decades. "You speak to them like you would to a person," says one mahout from atop his animal. "You say things like, 'You have to be a good girl. Now, why are you doing this?' After all, I brought up this elephant."

Nuon talks about a fellow mahout who a few years ago was saved from death by his sharp-witted animal which saw a massive tree keeling over on top of them and parried it aside with its trunk. On the other hand, Nuon has some scars to show from the time he was solicitously examining a bruise on one tusker's shoulder and in return got a kick that knocked him unconscious. But Nuon forgave the animal.

To mahouts, Nuon says, elephants mean money and a profession of some status. Most of the mahouts also ascribe "spiritual power" to the animal, weaving together strands of animist lore, Buddhism and Hindu mythology.

When an elephant leaves its mother to begin training, a spirit man comes to murmur incantations and pour over the baby's head and tail a potion of water and slightly burnt pods of the acacia tree. When a mahout leaves the school and his elephant, the acacia-flavored water is poured over the animal's head while the mahout confesses the wrongs he has committed against his beast during their years of training: unnecessary beating and abusive language.

About four years ago, says Chaum, the school's director, half a dozen mahouts at the training center used to sing and play instruments "to cheer up the baby elephants" after training. The animals, he says, would sway from side to side and saw the air gently with their trunks while the bamboo flutes wailed and the oriental violins struck up a tune. The musical mahouts have since moved on, so there is no music for the elephants anymore.

Some observers of the bonds between men and elephants in Asia note the striking parallels between the ages of both. Each starts school between the ages of three and five. The elephants can take on only light work between graduation and the age of about 15. The mature man and elephant put out their greatest energies between the mid-teens and late 30s, waning slowly through the 50s. Retirement for both elephants and the mahouts of the forest organization falls at 60.

The retired elephant, like the expectant mother, returns to the school. A mahout is assigned to do nothing but take care of its needs. Today, there are two retired elephants at the school, both in their late sixties. They live in the forest, and the mahouts visit them every two or three days. The two feed themselves, wandering through the forest along with deer, wild boar, squirrels, snakes and others. They will sleep three to four hours during the course of a day, laying down on their flanks, yawning and snoring like humans. Without the schedule of human-ordered work, they may well live up to their description as animals whose life is one continuous meal and sensual pleasure.

The mahouts look after the retired elephants' skin sores and other hurts. They bring bundled sticks of sugar cane, bananas and the elephants' favorite: sweet-sour pods of tamarind, wrapped up in a ball with a globule of salt inside. The mahouts continue coming into the jungle until one day their charges sink down to the forest floor for their last, long sleep. ∎

Denis D. Gray is Bangkok bureau chief for the Associated Press. AP photographer Neal Ulevich, also based in Bangkok, won a Pulitzer Prize last year for a series of photographs on the bloody suppression of a leftist demonstration in Bangkok.

Phantoms of the Prairie

Indians attributed supernatural powers to the explosive pronghorn antelope, and its recent comeback has been little short of miraculous

BY MICHAEL TENNESEN

AS TWO MEN picked their way across the windswept plateau of Oregon's Hart Mountain one day last winter, a fidgeting herd of some 30 pronghorn antelope watched them intently. The restless animals had long since stopped grazing and now their vivid white rump patches flared out dramatically, signaling danger. Horned heads shifted from side to side as enormous eyes strained to focus on the approaching threat. All at once, a doe bolted forward and, with that, the entire herd exploded into a spectacular stampede.

Mouths agape, the bounding creatures sucked in huge volumes of air through enlarged windpipes to nourish their oversized lungs and hearts. Effortlessly, they skimmed across the rough lava beds like desert wraiths, cruising along at 40 mph with ample speed to spare. Then, when the intruders did not test their quickness, the pronghorns glided to a halt and casually resumed nibbling the shrubs and plants around them.

Anyone who has ever chased a wild pronghorn knows how superbly the animal is engineered for escape. The antelope's speed, endurance and powerful eyes helped it become one of the most numerous mammals on the North American continent. With the settling of the West in the 1880s, the pronghorn population plummeted almost out of sight. But now *Antilo-*

capra americana americana and its cousin, *Antilocapra americana oregona*, are once again racing throughout the western plains in healthy herds, thanks to their physical grit and to growing public appreciation of their habitat needs. So great has the pronghorn's revival been, in fact, that today, it is one of the West's major game animals.

In 1818, after the explorers Lewis and Clark had collected the first specimen of pronghorn, scientists gave it the name *Antilocapra*, which means "antelope-goat." But actually, the pronghorn is neither a goat nor a true antelope. It derives from a separate family of animals that has been evolving on this continent for 20 million years. Unlike the deer, the moose, the bighorn and the bison, which all migrated from Asia, the pronghorn is a true native American.

In the 1930s, naturalist Thomas Seton estimated that there may have been 40 million pronghorn west of the Mississippi before the coming of the white man. They evolved on the open ranges, where their stamina developed in grueling elimination bouts with wolf packs. In speed, the prong-

A dramatic warning system enables pronghorn to flash danger signals to each other more than a mile away. When threatened, the sharp-eyed animals flare their white rump patches, triggering a chain reaction that passes quickly from group to group.

Photograph by David Muench

horn is reputed to be second only to the cheetah. Bucks average 115 pounds (does weigh around 90) and their hearts are twice the size of a sheep's. Their windpipes are fully five inches around to supply all the oxygen needed for their remarkable bursts of speed.

In his book, *The Pronghorn Antelope,* Oregon wildlife researcher Arthur Einarsen describes how one of the creatures picked a race with his car — something pronghorn frequently do. The buck matched the speed of the auto for a while and then darted directly in front of it "as the speedometer registered 61 mph."

Another of the pronghorn's primary defenses is its extraordinary vision. Its eyeball is as large as that of a horse (which weighs about eight times as much), measuring 1½ inches across and set in the socket so that the antelope has a wide-angle view. These eyes are said to be equal to the human eye aided by seven-power binoculars. The antelope's glistening white rump patch, when flared out as a warning signal, reportedly can be seen by other antelopes miles away. Says Pete Carter, director of the Hart Mountain and Sheldon National Antelope Refuges in Oregon and Nevada: "Many times I'll be out searching for pronghorn on the refuges and spot one that you can't see with your naked eye, but you can with your binoculars. And there he'll be, staring right back at me."

The pronghorn's elusiveness so impressed the American Indians that they often vested the creature with supernatural powers. To them, it was the phantom of the prairie. The Apache considered it a monster, while the Hopi thought it was a medicine animal and prayed to it for rain. One legend tells how the antelope was created by a Blackfoot god called the "Old Man." He made the antelope out of dirt in the mountains but, when he turned it loose, the pronghorn ran too fast, falling and injuring itself on the rugged terrain. So, the Old Man took the antelope down to the open plains, where it was free to run without harm.

That freedom ended abruptly when the white settlers arrived with their long rifles. Commercial hunters found

To escape from predators, a young pronghorn will dash at blinding speed for several hundred yards and then, suddenly, drop out of sight into grass. There, the fawn remains motionless for hours if necessary.

186

Photograph by John D. Fandek

Photographs by Bill McRae

a ready market for pronghorn meat in the West's new cities, and in the 1850s pronghorn steaks sold for 25¢ a pound in San Francisco. During a drought in 1859, one trapper built a blind near a water hole and shot some 5,000 pronghorn, taking only the hides and leaving the carcasses to rot. Farmers and ranchers plowed under the antelope's feeding grounds and began an indiscriminate slaughter of the herds, which they feared would compete with their livestock. By the turn of the century, the pronghorn had dwindled to less than one percent of the original vast herds and a government survey in 1910 showed only about 15,000 remaining in the U.S. The pronghorn was well on the road to extinction.

In the 1920s, however, public attitudes toward wildlife conservation started to change and state game agencies throughout the West — operating with greatly increased budgets — began putting full-time wardens into the field. Soon, hunting laws that had been on the books for years were being enforced and the wholesale, year-round slaughter of the pronghorn was brought to a sudden halt. By the late 1920s, homesteading also declined throughout much of the animal's range, signaling the return of short-grass prairies to many areas.

The results were dramatic. Pronghorn increased over 1,000 percent, to about 435,000 today, and they are now hunted in controlled seasons in 14 western states. Close to one million pronghorn have been harvested

in the last 50 years, according to Jim Yoakum, antelope biologist with the U.S. Bureau of Land Management (BLM) in Nevada, and the species remains strong and healthy.

Winter in northeastern California finds the pronghorn gathering into large herds and moving down to the lower elevations, where their forage is less hampered by deepening snows. Here and in neighboring areas of Nevada and Oregon, the antelope have made a recovery similar to the one that has transpired further east. Since 1950, Nevada, Oregon and California have held an annual convention to share census data as well as management techniques that have thus far proven successful.

It was in this part of the country that the only wildlife refuges principally for antelope were established in the 1930s on Hart Mountain in Oregon and on the Sheldon Range in Nevada. "At the turn of the century," says Jim Yoakum, "this area had some of the last remaining large herds of antelope in the nation."

Because the Hart Mountain and Sheldon country is high, semi-arid desert, it does not support the dense numbers of antelope that are maintained on the midwest grasslands. Here, the principal diet is sagebrush, forbs — small flowering plants such as daisies and sunflowers — and some young tender shoots of grass. Sagebrush and other shrub species become increasingly important on winter ranges when forbs and grasses are buried under snow.

During heavy snowfalls, pronghorn will often move to the wind-swept ridges free of snow where food lies uncovered. Jim Yoakum lived on the range with one herd during an entire winter. "It was often 25 degrees below zero, the coldest I've ever been in my life," he recalls. "Yet you could see the antelope up on these ridges frolicking and playing." In the high desert, wind can reduce temperatures on these ridges to -50 degrees F.

Fawning in the high desert begins in May, when does seek out ancestral kidding grounds on the open range to give birth to their young. The fawns, usually twins, take their first steps within 30 minutes of birth and by the fourth day they can outrun a man. This rapid maturity is necessary, as fawning takes place just at the time that coyotes and bobcats are hard-pressed to supply food to their own hungry young.

In spring, the winter herds begin to break up into smaller bands as the leaders — usually does — direct the journey back to summer ranges. "The unfortunate habit of the animals to

The fastest mammal in North America, the pronghorn antelope has been clocked at speeds of more than 60 mph on the western plains (above). To supply all of the oxygen required for such high-powered bursts, the animal has an unusually large windpipe measuring fully five inches around.

Complementing its speed, the pronghorn has powerful, wide-set eyes (right) that enable the creature to look around without turning its head. The pronghorn can see as well as a man aided by seven-power binoculars — a radius of about two miles, in all.

follow a leader is actually a handicap," reports Arthur Einarsen, "since any menace is ignored so long as a group leader follows a definite path." In 1944, one apparently frightened herd of 100 antelope followed such a leader to their death over a 45 foot cliff in Wyoming.

Another weakness of the pronghorn is its tremendous curiosity. By waving cloth flags, early antelope hunters could often tempt the animals in for a close shot. I had a chance to test this legendary curiosity while photographing a group of pronghorn on their winter range near Goose Lake, Oregon. Holding a camera on a tripod above my head, I was able to approach the animals by duck walking low to the ground and proceeding in a zigzag pattern, making several 360 degree turns when I noticed they were watching. I finally set the camera down and the group that had eluded me all morning approached to within 30 feet to see what this strange creature was doing on their territory.

Late summer signals the approach of the mating season. Restless males start herding does, who are fertile after their first year, into harems, and young bucks challenge older males to ritualistic battles. These short fights begin as two contending animals approach each other cautiously until horns meet. Vigorous pushing and shoving follows until the weaker of the two breaks contact and retreats. The prong on the antelope's horn catches the opponent's prong or horn and prevents gouging wounds to the head or neck, though fights can be bloody and sometimes fatal.

Competition for rangeland between antelope and sheep, which have similar food needs, can be intense. With the decline of the sheep industry in some areas, there has been a corollary increase in antelope. Interestingly enough, competition with cattle — which have different feeding habits — is not a major factor. In recent years, with antelope populations stabilizing throughout many parts of the West, management efforts have been aimed at improving the quality of existing rangeland.

Livestock fences are sometimes a barrier to antelope movements. Pronghorn can broadjump 25 feet but they have problems with fences only four feet high. One observer reports that antelope racing along at 40 mph have been seen bursting right through four-strand wire fences without hesi-

Undeterred by sub-zero temperatures, a pronghorn bounds across a snow-covered plain in Wyoming. Pronghorn are often forced onto high, wind-swept ridges to reach the forage still uncovered by snow.

tation, "leaving a cloud of hair floating in the wind." Studies in Wyoming have determined, however, that most antelope would rather go under a fence than over or through it, and so a minimum distance under the lower strand of livestock fences is now required in areas where antelope migrate.

Water is a critical factor in any good antelope habitat, especially in the high deserts of Oregon, Nevada, and California, where summer drought conditions prevail. There, wildlife agencies are continuously involved in water-hole projects, digging troughs in intermittent lake beds and erecting small dams to hold back water seepages and collect moisture for the dry seasons.

One improbable tool for improving antelope range is an erstwhile enemy: fire. When sagebrush grows to heights above 25 inches, antelope often leave because of decreased visibility. "Burning eliminates the big sage," explains Pete Carter. "Fire is one of the best wildlife management tools available."

Several sagebrush elimination projects were undertaken at Pony Springs in eastern Nevada, where sagebrush was either plowed or "chained," a practice in which tractors drag a large anchor chain over the terrain, knocking down shrubs but leaving forbs and grasses. This vegetation restoration program has led to increased antelope use in that area.

"Unfortunately, we now have more acres going out of antelope habitat than are being improved," warns the BLM's Jim Yoakum. Yoakum also has a stern warning for the future: "The biggest problem coming to antelope rangelands is strip mining."

Though future growth may be uncertain, the present antelope situation is clearly far better than existed at the turn of the century. The return of pronghorn herds "must rate as one of the principal achievements of wildlife management to date," says Joe Van Wormer in his book *The World of the Pronghorn*. Right now, the phantom of the prairie is no longer in danger of becoming a real ghost. ∎

Californian Michael Tennesen is a frequent contributor to NATIONAL WILDLIFE.

They may not be as glamorous
as the big felines, but
these creatures are far more
adaptable to man

The small cats of Africa

BY NORMAN MYERS

THINK OF CATS in Africa and you inevitably think of the lion, the leopard and the cheetah. You may think, too, of how rapidly they are losing numbers. The lion, poisoned far and wide by African stockmen who covet its savannah habitat for rangeland, is disappearing faster than the zebra or the wildebeest. The cheetah's total could drop below 10,000 by 1980. The leopard, more resourceful and with better prospects than the other two, is plagued by fur poachers.

But Africa contains many more cats besides the big three. In all, there are 20 small felids and catlike viverrids on the continent and several of them are not faring badly. A few could even be doing something generally unheard of for wildlife in modern Africa: holding their own. In some cases, moreover, this happy circumstance has actually been occasioned in part by man's disruption of natural environments.

In their own way, these small cats — serval, caracal, civet, genet and others — are every bit as interesting as the big ones. Pound for pound, they do not carry so much muscle as a lion, and certainly not as much as a

Photograph by Bruce Coleman, Inc.

leopard. But they often exhibit more stealth, cunning and ferocity.

Although these small cats probably inhabit Africa by the hundreds of thousands, the only place where they can reliably be seen is in a zoo. In the wild, they are extremely secretive. After 18 years in Africa, I have seen a serval twice, a civet once and the rest not at all. But I have come across lions and cheetah hundreds of times.

The explanation: a small cat knows how to conceal itself in the smallest bush or a mere clump of grass. Without this capacity, it would go hungry. Because it does not have massive power to overwhelm its prey, it must depend more on cunning and stalking skill. Also, the small cat's slight build usually means it cannot kill a prey large enough to last for several meals, as a big cat can. Whereas a lion or a leopard usually hunts creatures its own size or bigger, a small cat must

Not a true cat, or felid, the African civet (left) is a viverrid, which puts it in the same family as the mongoose. But its behavior is catlike, and it is popularly considered a cat.

An ancestor of domestic felids, the wild cat (right) feeds on birds, small rodents, insects and lizards. It also eats fresh grass.

Photograph by Clem Haagner (Bruce Coleman, Inc.)

be content with prey half its size. That means it must hunt more often and with greater care, which may explain that hair-trigger temperament and restless energy.

A small cat is a solitary animal, not inclined toward the social systems of lions with their prides. It cannot, therefore, rely on others of its kind to participate in a joint attack that allows larger prey to be dragged down. This places a premium on surprise and close-quarters attack. As a result, small cats have not evolved stamina for an extended chase. A lion or a leopard can sprint for 50 yards or more, but a small cat likes to seize its prey within just a stride or two — or it must look elsewhere. Furthermore, a small cat hunts small creatures that are themselves not built for defense through flight. Rather, the cats tend to depend on concealment by camouflage and inconspicuous movement.

For all of these reasons, it is inaccurate to say that small cats are "like the big cats only smaller." They are a whole different story. A serval is as specialized in its lifestyle as a lion. It is also as distinct from a caracal, a genet or a civet as a leopard is from a cheetah. Because of their different ecology and behavior patterns, small cats tend to live in a different set of habitats from lions, leopard and cheetah. They favor border zones between two environments — where forest gives way to savannah, for example, or where grassland merges into a wooded landscape. Here, they can meet their own distinctive food needs with prey from two ecological worlds. In these localities, they also find conditions where they can display their own hunting tactics and killing methods. Their mode of survival is a world apart from that of the large cats.

IN FACT, there is not one main group of small cats but two. The felids or "true cats" include the serval, caracal, the golden cat, the sand cat, the wild cat, the black-footed cat and the swamp cat. These animals purr rather than roar. In bright light, their eyes narrow to vertical slits, whereas a large cat's contract to pinpoints. They have a fairly large head to house the well-developed brain, and they have rounded faces to accom-

An adept hunter, the caracal can combine stealth with exceptional violence, killing prey almost its own size. Long hind legs enable it to jump high in the air after birds.

195

modate the muscles for their short, powerful jaws.

By contrast, the viverrids — civets and genets — aren't really cats at all. They are popularly known as "cats," though, and they are catlike in behavior and ecology. About the size of domestic cats, viverrids have relatively small heads and long pointed snouts; they are built more like weasels than cats. In addition to the civet and the genet, this family also includes the mongoose.

Whereas the large cats are found in many parts of Africa, certain small ones are much more restricted in their range. The sand cat, for instance, is found only in Algeria and Libya north of the Sahara, and in parts of Niger and Sudan in the desert's southern fringe. It also extends to the Middle East and even to southwestern parts of the Soviet Union. The black-footed cat is confined to dry sectors of southern Africa, notably the Kalahari. The golden cat inhabits forests of the west African coast from Sierra Leone to Cameroon, parts of Congo, and of Uganda and western Kenya. The swamp cat is confined to the lower Nile valley and the river's delta, where it favors marshy areas.

Others, however, range over large tracts of Africa. The serval and the caracal are found in north Africa and throughout virtually all the sub-Sahara region except forests. The caracal is also found in Asia as far as India. Much the same broad distribution applies to the wild cat, though it keeps to savannahs more than the other two, and it extends to southern Asia. The main species of civet extends from Senegal on the Atlantic Ocean to Somalia on the Indian Ocean, and then right down to Zululand and northern parts of Botswana and Namibia. The palm civet is widespread in forests of West and Central Africa from Guinea to Mozambique, while the aquatic civet lives near streams and rivers in dense forests. The small-spotted genet is found in virtually all of Africa outside the deserts and the large-spotted genet is fairly common in most of the continent south of the Sahara. The other

Adapted to semi-desert areas fringing the Sahara, the sand cat (kitten, right) has hairy feet that protect the animal from hot sand.

Weasel-like in build, the large spotted genet (far right) is a viverrid. Although it operates readily in trees, it prefers rodents to birds and eats bushbabies and lizards.

196

eight species of genet, however, are confined to just a few countries or parts of only one.

This extraordinary spectrum of small cat species, habitats and ranges represents a far greater variety than the large cats can muster. But for all that, small cats do not enjoy the super-star status of the big boys. They pose no special physical threat to man and consequently they do not generate the awe and fear that invests the larger, more dangerous beasts with an aura of glamour. This is one reason why small cats have not received much attention from wildlife researchers and why so little is really known about them. It is clear, though, that within their broad spectrum, each small cat occupies a very special niche.

AT HOME in several million square miles of Africa, the serval avoids damp swampy places and waterless plains. Along the water courses of grasslands, in woodlands and in thickets, it thrives. In Kenya it is found in the hot coastal belt, in the thornscrub of Tsavo Park, in savannah woodlands, and in mountain areas. A good place to look for it is in the Aberdare National Park, above the montane forest belt on the moorlands. There,

you can sight it with only a fair degree of luck and a good pair of eyes. Often, it appears in melanistic, or darkened form, as do the leopard and other cats living in moderately moist conditions about 10,000 feet.

I once came across a serval among tangled vegetation in Ngorongoro Crater. It was late in the afternoon when I suddenly spotted something that looked like a dark tuft swiveling around above the foot-high grass. My binoculars showed it was a serval's ear. After looking for awhile, I saw the other ear pointed almost straight-forward, horizontally. A quarter of an hour later, I saw the serval's head peek out above the grass for a cautious look around.

This sequence is typical of the serval's hunting technique: it tries not to show itself and moves with extreme care and quiet. This is, of course, in keeping with its life as a solitary small-sized cat. A serval, finding much vegetation as tall if not taller than itself, cannot locate the exact site of its potential prey by sight or smell, so it depends on some of the most highly-tuned hearing apparatus in Africa.

It turned out that the serval I was watching had a mole rat in its jaws. This is typical of the serval's prey,

Photograph by Tom McHugh (Photo Researchers, Inc.)

which generally amounts to small rodents, ground birds and the like. But, although it weighs no more than 35 pounds and is delicate in build, it can take birds as large as a guineafowl, weighing perhaps ten pounds. Sometimes it tackles hares, duikers, oribis, even gazelle and impala fawns. It does not even ignore lizards and fish.

Like the cheetah, the serval will hunt during the day, especially during the early morning hours and late afternoon. Its principal hunting period, though, seems to be nighttime. During the hot hours of the middle of the day, it prefers to lie up in an abandoned aardvark or antbear burrow, or under a shady bush.

EVEN MORE CATLIKE than the serval, the caracal, with its supple body and swift reflexes, is an adept hunter in a broad range of situations. A caracal at work was once observed by an ecologist friend of mine, who used a movie camera to record the action. He came across it as it was peering around a bush at a flock of ground-feeding plovers. When the cat launched its attack from a few yards away, the birds scattered, becoming airborne almost immediately. The caracal caught two of them in the course of a single leap, and the movie film reveals how it changed course in mid-air to strike the second one. Not even a leopard could match that feat.

Both the caracal and the serval show a similar technique for handling the prey after the first strike. Possibly from fear of retaliatory attack, they use a single sweeping movement to grasp the prey and then to fling it a fair distance before it can strike back. While the surprised creature is recovering, they pounce on it again, only to repeat the entire tactic. After several such attacks, which leave the prey thoroughly disoriented and battered, the cat feels assured enough to carry through a final assault. It grips the prey with teeth and claws, until the last struggles fade away.

Both cats combine surprise with exceptional violence for such small animals. So successful is this strategy for the caracal that it can kill a small antelope such as a dikdik, a prey nearer its own size than the customary birds, rats and ground squirrels.

THE COMMON GENET inhabits thick-grown vegetation, whether grass tangles, thickets, scrub growths, riverside vegetation. Unlike the serval and caracal, it also prospers in the forest. About the size of a small domestic cat, it is more weasel-like in build than the other two creatures, in accord with the pattern of viverrids. But its body and snout are longer, and the genet appears something like the progenitor of Africa's cats is supposed to have looked 20 million years ago. It prefers rodents to birds though it operates in trees more readily than the serval or caracal and feeds a lot off bushbabies and lizards. It is even partial to scorpions.

The genet has its own special killing technique. After seizing its prey, it rolls over onto its side, so that it can grasp the animal with hind legs as well as forelegs and jaws.

MUCH MORE AT HOME in undergrowth is the civet. About the size of a cocker spaniel, and four times the weight of a genet, it operates only at ground level. It likes small mammals, birds, reptiles, frogs, toads, fish, crabs, insects and millipedes. Because it also relishes carrion and favors poultry, it prowls around villages where its crafty nature allows it to thrive without much retribution from man. It even eats plants.

The civet does not seem able to deliver a *coup de grace* to its prey with a bite behind the head in the precise style of many cats. Instead, it applies the "catch and throw away" tactic,

followed by an energetic shaking of its prey. This process can be violent enough to break the spine of a rat.

The civet has perhaps developed this strategy through experience with snakes, which are a frequent item in its diet. On releasing a snake, the civet leaps high in the air, presumably a trick learned to avoid snakes.

The palm civet is a different kind of creature altogether. It lives almost entirely off fruits. It does not need to practice the opportunist hunting of the more common civet and it gets by with a less catlike build. The aquatic civet, which has webbed feet, is somewhat in between the other two. It lives mostly off fish.

THE AFRICAN WILD CAT is an ancestor of the domestic cat. Like others of the felid group, its staple prey is small rodents and birds as large as a hen, with insects and lizards thrown in for good measure. Curiously, the wild cat sometimes seems to need plenty of fresh grass in its diet to keep in good condition.

Of the other felids, the black-footed cat is especially adept at the neck-aimed "kill bite." The golden cat, a tabby-sized creature which varies from golden brown to very dark grey, is particularly adapted to tree climbing. The sand cat, which is at home in semi-desert areas, has hairy feet to protect it against hot sand. And finally, the swamp cat, larger than a wild cat, has unusually long legs and a short tail.

WHAT OF THE FUTURE for small cats in developing Africa? The leopard and cheetah have been persecuted for their skins to meet the demands of the international fur trade and all big cats are encountering increased conflict with Africa's burgeoning livestock industry. Fortunately, however, stockmen of emergent Africa are less likely to become incensed at small cats. Because of their different diets, small cats are pretty much indifferent to most changes in savannah lands.

Even so the serval, genet, civet and golden cat are blessed (or cursed) with fine markings on their fur that make them prime targets of poachers. The serval especially possesses a splendid pelt. While conducting a recent survey throughout Africa south of the Sahara to assess the status of the leopard and cheetah, I found piles of serval skins in one bazaar after an-

Photograph by G. D. Plage (Bruce Coleman, Inc.)

other. In the back streets of Omdurman in Sudan, where cheetah and leopard skins are sold at the risk of lengthy imprisonment, serval skins can be found by the dozens. In Dakar in Senegal, Douala in Cameroon, Monrovia in Liberia, Niamey in Niger, Timbuktu in Mali, Windhoek in Namibia, Livingstone in Zambia, and Addis Ababa in Ethiopia, the same story holds.

Several African governments are now in the process of ratifying the Endangered Species Convention, which bans international trade in skins of leopards and cheetah. As African leaders step up the pressure on dealers in leopard and cheetah skins, a ready substitute is being found, unfortunately, in serval skins.

Conceivably, serval stocks in the wild could sustain this exploitation — albeit for tasteless purposes and often through illegitimate dealings — for quite a few years. On the other hand, widespread pressure could rapidly depress populations, until some are badly depleted and some disappear altogether.

But what about the greater threat that faces most African wildlife in the long run — the loss of something to eat and somewhere to live? Since small cats live off birds, rodents, lizards, frogs, snakes and plants, they will probably be little affected by the decline of larger prey animals. In fact, when grasslands are planted in corn and when forests are chopped

A tabby-sized felid, the golden cat (left) is adapted to tree-climbing and inhabits forests of the west African coast. Sometimes its beautiful spotted skin is gray or black.

Large ears are part of the highly tuned hearing apparatus that helps the serval (right) locate the exact location of its prey.

down, the new forms of vegetation may actually supply better conditions for seed-eating birds, rats and the other small creatures that make up the diet for small cats. Moreover, civets, genets and probably other small cats can produce a litter of four to six young ones twice a year, compared to the lion and cheetah, which may not produce a litter of that size more than once every two years.

There is a distinct possibility that, as peasant villages spread across more and more of Africa's wild lands, the continent's small cats will eventually wear the "varmint" label American stockmen have pinned on the coyote. There will be more kid goats for civets to maraud, more poultry for serval and caracal to steal, more eggs for genets to eat.

THE SECRETIVE, stealthy lifestyle of small cats also serves them well when man intrudes on their habitats. Sonia Jeffrey, a British ecologist who has lived in back-country areas of Ghana and Liberia for eight years, tells many a tale of how civets — with their crafty and persistent hunting methods — can take chickens night after night, prowling near villages without being caught.

This account and others like it confirm the impression I have derived from "cat talk" with hundreds of experienced researchers, game wardens, farmers, scientists, park rangers and trappers. It seems that small cats are better adapted than many other animals to the changes overtaking Africa. But this is only an impression.

In the final analysis, we know all too little about small cats to draw any firm conclusions. There are scarcely a dozen detailed accounts on all of the small cats put together and until more information is available, I shall harbor the hope that, if Africa ever degenerates into a land where the most prevalent creatures are sparrows, rats and poultry, at least some of these mini-cats will fare far better than most other forms of wildlife. ■

Author, ecologist and photographer Norman Myers is a Roving Editor for INTERNATIONAL WILDLIFE.

Photograph by Norman Myers (Bruce Coleman, Inc.)

PHOTOGRAPHS BY STEPHEN J. KRASEMANN

You can't sneak up on a marmot

Early French trappers called this overgrown woodchuck *le siffleur* . . . the whistler

BY JEROME J. KNAP

I watched the grizzly bear tear at the ground with its long claws, trying to uncover the cluster of burrows that housed a marmot colony. The enraged animal moved with incredible speed, huffing angrily as it ripped first at one burrow and then at another. Finally, it gave up in frustration. The hoary marmot had won again.

Over the years, I have seen the evidence of similar scenarios recorded in torn turf and overturned rocks, and I once watched the canny creatures outfox a coyote. At each little victory, my affection for the hoary marmot has grown.

The coyote incident happened in northern British Columbia, where I was inspecting a distant basin in hopes of seeing some Stone sheep. My binoculars settled on a colony of hoary marmots, busily stocking up on grass in anticipation of the long winter's rest. It was late August, and the fattened adults weighed up to 30 pounds or so. As always, I was amused by the scurrying antics of these animal clowns, which resemble overgrown woodchucks.

The animals' muzzles and heads were black, as were the tips of their brown bushy tails. The hoary marmot's inelegant name comes from the grizzled mane that covers its shoulders and foreback, but its scientific name, *Marmota caligata*, refers to its black feet, which resemble the caligae, or black leather footwear of the ancient Roman legions. I much prefer the French Canadian trappers' name: *le siffleur*, the whistler — a reference to the marmot's high-pitched cry of alarm. It also can hiss, meow and — when angry — click its teeth.

The colony I was watching darted back and forth, gorging on the succulent grass. Suddenly, a sentry sounded the piercing warning and the marmots scampered for their den openings. There they waited, ready to disappear if the danger materialized, be it fox, wolf, golden eagle or bear.

I looked around the area and, sure enough, there was a coyote. The wily predator apparently knew the jig was up, for he gave up his stalking, hunched his shoulders and trotted off to look elsewhere for a meal.

Score another one for the hoary marmot, which most people take for granted despite its uncommon knack for survival. Over most of its mountainous range in the western U.S. and Canada, far more glamorous mammals — wild sheep, goats and grizzlies

A hoary marmot checks for danger at the entrance of its burrow. One of six marmot species in North America, it gets its name from a mane covering its foreback.

— seem to have a monopoly on the public's attention. The irony is that the marmot's curious ways may hold the clue to a better understanding of these and other creatures.

"In many ways, marmots are ideal subjects for research on social behavior and social biology among mammals," says David Barash, associate professor of psychology and zoology at the University of Washington. Barash says that marmot colonies — unlike troops of primates, packs of wolves and other animal groups that have been studied extensively — are easy to locate, stay in one place and are easily observed because they are active only during daylight hours. He claims that the complex social structure of the colonies offers untapped opportunities to study many aspects of animal behavior.

Anyone who has watched a marmot colony will know instantly what the professor means. For example, when two animals meet, they may sniff at one another. Sometimes they nuzzle almost as though kissing or even bat each other with their paws in a sort of mutual backslapping.

Impromptu wrestling matches happen frequently, and the other members of the colony will stop to watch. Often, the combatants nip at the fur on each other's throats. Occasionally, the opponents lock teeth and hit with their paws. If the contest takes place at the top of a hill, the two will sometimes roll and tumble to the bottom, break their hold and then dash back up to start all over again.

To Barash, who has studied a close relative of the hoary, the Olympic marmot, all this greeting and wrestling are much more than fun and games: "The greeting is a sort of social glue. It is how the marmots recognize other members of their colony." The wrestling, he says, is a form of aggressive behavior that establishes which male is dominant within the colony.

Another marmot-watcher, David Gray, assistant curator of vertebrate ethology at the National Museum of Natural Sciences in Ottawa, has described a typical summer's day in the life of a hoary marmot in southern British Columbia. They wait until sunrise to leave their burrows, Gray says, and after warming themselves in the sun for a short time, they begin to feed, often interrupting their eating to wander about. In warm weather, they will go down into their burrows for a few hours at midday. In the afternoon, they may sun themselves a little, walk around a bit and then begin their afternoon feeding, which is never as intense as in the morning. By six o'clock or so, some of the members of the colony will have retired for the night. The rest of the colony is quiet before sundown.

Hoary marmots generally make their homes above timberline in rolling mountain meadows and rocky taluses. In the Canadian Rockies of Alberta, they are seldom found below 6,500 feet, though in the Yukon they sometimes live as low as 2,500 feet, and in Alaska they move right down to sea level. The animal's range extends northward from Montana and Washington and covers some of the wildest country on the continent.

Of the six species of marmot in North America, the widely distributed yellow-bellied marmot and the woodchuck have been studied the most. The others are: Brower's marmot (found in the Brooks Range of Alberta), Vancouver Island and Olympic marmots, and the hoary.

It is probable that the hoary marmot shares some of the traits of its closest relative, the Olympic marmot of the Olympic Peninsula, which breeds only every second year (the woodchuck and yellow-bellied mar-

203

mot breed annually). Litters average four or five, and the young remain with their mothers for two years. Growth is relatively slow.

The marmot is a vegetarian, eating succulent roots, wild onions, grasses, sedges, lichens and alpine flowers such as rosy douglasia, miner's lettuce and mountain phlox. When a colony feeds, its members are always ready to warn of nearby danger.

The hoary marmot never has been hunted much by white men, though the voyageurs and mountain men are said to have relished its meat. The Indians often hunted it for its soft hide, which was prized for sleeping robes, and the meat was also eaten. Indians who still live traditional life-styles hunt marmots today. I once met two Indian boys hunting marmots in the Yukon. They told me that whistlers are better to eat than rabbits.

By and large, though, the hoary marmot has little to fear from modern man. There is no indication that populations of the comical creature have declined, and that's a good thing. Without its antics and whistling cry, mountain trails would be poorer, lonelier places. ∎

Jerome Knap — biologist, writer and former guide, trapper and forester — lives in Guelph, Ontario.

A mouthful of grass means spring cleaning and a refurbished nest, above, but grass is also a common food on the rodent's menu, right.

Combatants in wrestling matches nip at each other's fur and sometimes lock teeth or hit with their paws in a common activity that helps to establish male dominance in the social structure of a colony.

Perfectly adapted for nighttime hunting, this fishing bat locates its prey with sonar and snares it with oversized claws.

IN CELEBRATION OF
BATS

A
hidden world
comes to life

A S DARKNESS falls, nearly a thousand kinds of flying mammals take to the air. They are bats, and using ultrasonic signals, they flit across the evening skies, navigating and communicating, masters of the night.

Bats are found worldwide—except for extreme desert and polar regions—and they account for almost one-fourth of all mammal species. But they reproduce slowly and usually live where they are easily disturbed. That makes them vulnerable. Indeed, their populations are declining rapidly and some are already extinct.

That this can happen without outcry is due in large part to human ignorance. Few people realize that bats are gentle, intelligent, and frequently beneficial animals. The following facts and photographs celebrate these "creatures of darkness" and are intended to shed new light on their remarkable lives.

STORY AND PHOTOGRAPHS BY MERLIN D. TUTTLE

Big ears and a complex nose-leaf, or flap, form a precise echolocation system in this Asian horseshoe bat (above).

This New World leaf-nosed bat (left) is well named; flaps of skin on its face help to direct its sonar.

A fishing bat makes a meal of a minnow (right). Its fur has been bleached by ammonia fumes from the guano in its roost.

The frog-eating bat (far right) of tropical Latin America finds its prey with "normal" hearing—rather than sonar—and catches it with its mouth.

WHAT THEY ARE

OLD MASTERS OF THE SKY: Bats are the only truly flying mammals, and their independent wing and tail movements permit extraordinary aerodynamic maneuvers, unparalleled by any bird or flying machine. The bones of a bat's wing are essentially the same as those in human arms and hands—bat fingers are simply extended and connected by a double membrane of skin.

Thus the name of the scientific grouping—the order Chiroptera, meaning "hand-wing." They are divided into two suborders. The first consists only of flying foxes: these large bats of the Old World lack sonar in most cases and navigate by sight alone. The second consists of all other bats: these are distributed worldwide and possess echolocation, a type of sonar. In all, there are nearly 1,000 species. Both groups were already well developed 50 million years ago, and are highly sophisticated—bats are more closely related to people than they are to mice.

TINY BATS, HUGE BATS: The Kitti's hog-nosed bat of Thailand, weighing less than a penny, is the world's smallest mammal. In contrast, the largest bats—flying foxes—sometimes weigh more than two pounds, with wingspans of up to six feet—as wide as a small eagle.

SEEING WITH SOUND: Bats are famous for their sonar. They produce rapid pulses of sound at rates ranging from 20 to more than 500 per second; the energy in a little brown bat's echolocation call is as intense as that of an alarm of a smoke detector. Bats use sonar to perceive motion, distance, speed, trajectory, shape, texture and size of even a tiny mosquito. Some can detect and avoid obstacles no wider than a human hair. They also can track only their individual sonar, even in caves filled with millions of echolocating bats. Their abilities far surpass our current understanding.

FACES THAT NAVIGATE: Faces that seem ugly to the uninitiated are beautiful to scientists who study bat sonar. Some bats have huge ears, nose-leaves, flaps or other appendages with strange shapes. All are critically important to navigation. For instance, leaf-nosed bats use their nose-leaves, a flap of skin around the nose, to "beam" ultrasonic transmissions, and leaf-lipped bats have leaflike skin around their mouths which can be puckered into a megaphone.

WHAT THEY EAT

FROM MOSQUITOES TO MICE: Bats have adapted to a variety of diets, but they first ate insects, and most, about 70 percent, still do. Bats are the only major predators of night-flying insects. A single gray bat may eat 3,000 or more during a night's feeding, and large bat colonies consume countless billions. For example, free-tailed bats from a single cave in Texas may eat 500,000 pounds or more of small insects nightly.

Nearly all of the remaining bats prefer fruit or nectar, although about 13 percent are carnivorous, feeding on small fish, lizards, frogs, mice, and birds; some insectivorous and carnivorous bats even eat such unlikely prey as scorpions. Finally, despite all the publicity, only one-third of one percent drink blood.

Because they eat insects, disperse seeds and pollinate flowers, most bats are beneficial to people. A few Old World fruit bats damage crops, but such problems are often exaggerated. These bats eat only strong-smelling ripe fruit, and seldom damage commercial crops, which are picked green for shipment; they are often blamed for damage actually caused by birds or rats. In Latin America, vampire bats can harm live-stock, but attempts to control them often have led to the destruction of millions of beneficial bats instead.

WARRING WITH PREY: Bats are efficient predators, and their prey have evolved specialized escape strategies. Many moths, green lacewings and other insects listen for bat sonar and attempt to escape with evasive flight patterns. Still others produce defensive sounds which may be used to warn of distastefulness or to confuse the bats' senses. Bats respond to these defenses by flying faster after their prey, using weaker echolocation signals to reduce the distance at which insects can detect them, or by "turning off" their sonar and simply listening for wing-beat sounds or mating calls produced by their prey.

Frogs, in turn, also have developed ways to hide from carnivorous bats that home in on their mating calls. Some frogs call only on bright, moonlit nights, when they can see bats coming in time to stop calling or evade them. Some call from thorn bushes, or from beside noisy waterfalls that apparently interfere with bat hearing. Others call only from large groups, playing a kind of frog Russian roulette.

WHERE THEY LIVE

FROM CAVES TO PALM FRONDS: Bats traditionally live in caves, and indeed some caves contain huge colonies. Cave-dwelling species may roost at densities of up to 300 per square foot, with as many as 40 million or more living in a single cave. Other bats prefer to live in cavelike hollow trees or buildings.

Less known are bats that roost in animal burrows, termite and bird nests, unfurling banana leaves, bamboo stalks or even spider webs. Several make their own homes by cutting palm fronds or other leaves to make "tents." Those that live in banana leaves use special suction-cup disks on their wings and feet to climb slick leaf surfaces. And the tiny bamboo bat enters bamboo through small beetle holes; it is reported to enter so fast that it avoids any visible break in flight.

FLYING BETWEEN HOMES: Most temperate-zone bats migrate in the spring and fall to find suitable hibernating caves or warmer climates. Many travel hundreds of miles; some European species may migrate up to a thousand miles one way.

Although some bats find their way home even when displaced hundreds of miles by researchers, no one knows for sure how they navigate. One species visually orients toward mountains miles away, but others are capable of migrating to traditional distant sites even when blinded.

Almost certainly, young bats must learn from older, experienced bats. For example, gray bats that breed and hibernate in a single cave often come from all directions to a place they have never seen. They could not be genetically programmed to go in one direction, and their sonar reaches only a few yards at best—yet these bats travel hundreds of miles and find cave entrances too small and obscure to be detected from more than a few feet away.

SURVIVING WINTER: Most bats survive food shortages or bad weather simply by turning down their metabolism. An active roosting bat may have a heart rate of 400 beats per minute (more than 1,000 when in flight) and a body temperature of 97 to 106 degrees F., but in hibernation, its heart rate may drop below 25 beats per minute and its body temperature may approach freezing. Although some bats, such as pipistrelles, easily freeze to death, hibernating big brown bats in Wisconsin were found unharmed with icicles hanging from their fur and body temperatures below 30 degrees F.

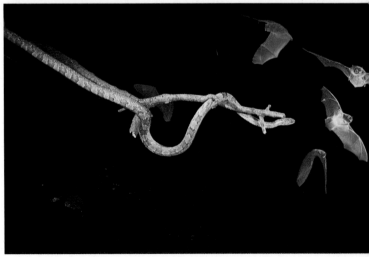

Latin America's tiny disk-winged bats (above), shown here at half life size, are named for the suction cups that hold them to their slippery roosts in unfurling leaves.

Three lesser short-nosed fruit bats (top left) have found a home under a palm frond in Thailand.

Cave-dwelling Rhoad's bats of the southeastern U.S. often fall victim to the gray rat snake, which waits outside the entrance for the evening exodus (bottom left). When a passing wing tip touches it, the snake strikes, constricting the unwary bat to death (bottom right).

A newborn gray bat (top) has strong clinging instincts; it can die in a fall from the cave roof.

Close quarters are a necessity for these two-week-old gray bats (left). This pocket in a cave ceiling traps their body heat, allowing them to conserve energy for growth.

Young bats may grow faster when different species roost together, profiting from each other's warmth. These short-tailed and large-eared bats (right) were photographed in a hollow tree in Panama. Short-tailed bats eat fruit, and large-eared bats eat insects, thus avoiding competition.

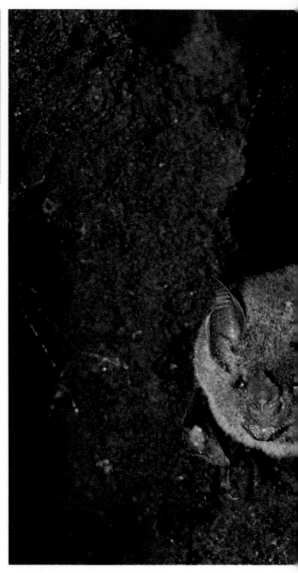

WHY WE FEAR THEM

BATS AND DISEASE: Sensational news and magazine stories often portray bats as vicious, filthy, and likely to attack and transmit diseases. Such stories are greatly exaggerated. Bats are actually gentle and meticulously clean; they seldom transmit disease or parasites to humans. Despite frequent claims that many bats are rabid, fewer than one half of one percent actually contract the disease, and even these rarely become aggressive. Any bat found on the ground or in an unusual place should be assumed to be sick. It should be avoided, as should any sick animal.

Only ten people in the United States and Canada have died of disease from bats in more than 30 years. In contrast, more people are killed annually, in the United States alone, by dog attacks, bee stings, power mowers or lightning.

SUPERNATURAL POWERS: From earliest times bats have had an aura of supernatural mystique. In China they are often seen not as omens of evil, but of good luck and happiness. Buddhists sometimes consider them sacred. The Chinese word *fu,* for bat, is also the name of the character meaning happiness, and thus the character for bat also stands for good

luck and happiness. The bat symbol has become highly conventionalized and is a favorite design on everything from bowls and lacquer boxes to emperor's thrones. Designs of five bats are symbolic of the five great blessings sought by all people: health, wealth, long life, good luck and a desire for virtue or tranquility.

According to Muhammadan legend, Christ personally created bats during a fast among secluded hills. No food could be eaten between sunrise and sunset, so He created bats to apprise Him of these times. In Central America, the bat god was a powerful deity of the Mayans. On the other hand, medieval Europeans associated bats with imps and devils, and in medieval portraits the Devil has bat wings and pointed ears.

Association of bats with the supernatural made them a favorite source of material for people who dealt in ancient magic. Bats were used as amulets or as ingredients in many magical concoctions that were said to induce sexual desire, to cure venomous bites, to aid in childbirth, to relieve rheumatism, to prevent baldness or to cast spells.

HOW THEY BREED

GETTING TOGETHER: Like people, bats use sexually attractive body language, odors and vocal exchanges to attract mates. In some species, elaborate courtships may ensure loyalty. In one case the female initiates the sex act when she enfolds her suitor with her wings. Intimacies include prolonged soft whining and mutual licking and grooming.

Not all bats are so romantic. Some keep harems, and others are simply promiscuous. Male sac-winged bats often guard a prime feeding site, permitting only females that mate with them to feed there. Male Jamaican fruit-eating bats defend their home—usually a hollow tree cavity where females roost—by chasing other males away. Male African hammer-headed bats prefer to gather in groups at select sites, where they sing and engage in communal displays. Females fly around among the singing males until a choice is made. Then they copulate and leave to become single parents.

Many bats that hibernate in caves mate in the fall. The males appear to be completely promiscuous, and will even copulate with sleeping females, continuing until they run out of energy.

DAY CARE: Most expectant mothers join nursery colonies where hundreds or even millions of them may congregate to rear young. In most species the mother raises only one offspring a year, and in all but a few species each mother recognizes and cares for her own. She is very solicitous, often protecting her young in her wing and quickly rescuing it if it falls or strays. Mother bats also may help one another. Some carry their young to safe hiding places in nearby foliage before leaving to find food.

In species that form large colonies, expectant fathers usually cooperate by simply leaving—sometimes living and feeding hundreds of miles away. This reduces competition for limited food resources near the nursery site. However, in a number of species that live in small groups or pairs, males may help with rearing young. In at least one case, adults apparently take turns guarding the young, and fathers bring food to the roost for their mates.

Young bats grow rapidly, and often fly in only three weeks after birth. At that time they can be found traveling with their mothers, apparently learning to hunt.

The structure of a flying fox (above) is revealed against the sky of Thailand.

Bats gradually are yielding their secrets to researchers like the author (left). He is shown here emerging from a tree in Panama, where he photographed the bats on the previous page.

Its cave-dwelling habits expose the endangered gray bat (right) to interference and death from unwitting explorers.

Bats are major pollinators of tropical plants; the big fruit-eating bat (far right) of Latin America will carry pollen to the next white shaving-brush flower.

HOW THEY HELP PEOPLE

GROCERY STORES AND BATS: Grocery stores would not be the same without bats. The fruit-eating species are nature's most important seed-dispersing mammals, and nectar bats pollinate more than 130 genera of tropical and subtropical trees and shrubs. The long list of valuable fruits, nuts, spices and derivatives from bat-adapted plants includes:

Peaches	Cashews
Plantains	Cloves
Bananas	Tequila
Palm sugar	Javanese long pepper
Breadfruit	Chicle latex for chewing gum
Mangoes	Manila and sisal fibers for rope
Guavas	Kapok fibers for life preservers and
Figs	surgical bandages
Carob	Balsa wood for crafts and other
Avocados	lumber for furniture

Many economically important plants depend upon bats. For example, durian trees of Southeast Asia are pollinated al-most entirely by a single species of bat. Two-and-a-half acres of these trees may produce fruit worth $10,000 or more annually, and regional sales of durian fruit total almost $90 million per year.

TONS OF FERTILIZER: In the southeastern United States as much as 100,000 tons of bat guano was once mined from a single cave. Guano, which contains nitrogen, continues to be a major source of fertilizer in underdeveloped countries, particularly in Southeast Asian countries such as Thailand and Sarawak.

MEDICAL BREAKTHROUGHS: Bats are extremely valuable in scientific research. Because they use highly sophisticated sonar, are exceptionally long-lived and disease-resistant, can survive unique environments, utilize sperm storage and delayed implantation, and have nearly transparent wing membranes, bats are increasingly important. For example, research on bats has contributed to development of navigational aids for the blind, new vaccines, artificial insemination and birth-control methods, drug testing and to studies of aging and space biology.

WHY THEY'RE IN TROUBLE

PERSECUTION BY PEOPLE: Despite their obvious value, bats are among the most relentlessly persecuted animals on Earth, and many of their populations are already dangerously low. Survival of these bats depends upon prompt education of the public. Sensational scare stories are dangerous both to bats and people—often fears lead to dangerous solutions to minor or nonexistent problems. Human homes are needlessly threatened with dangerous toxicants, for instance, and huge colonies of highly beneficial cave bats are wiped out in mass killings.

Only a relatively few caves can support bats and these are often of critical importance to a species. Hundreds of thousands or even millions of bats may migrate from surrounding areas to hibernate in a single suitable cave. These colonies are extremely vulnerable. Closure of one cave entrance, or repeated human disturbance, can eliminate whole populations in a very short time. Hibernating bats may lose 10 to 30 days worth of essential fat reserves from the passage of one cave explorer. Many young die when their mothers are disturbed at their summer nursery roosts—a disaster since,

for their size, bats are the slowest reproducing mammals.
LEGISLATING PROTECTION: Bats are legally protected in all European countries and in Russia. In fact, thousands of bat "houses" have been placed in national forests throughout Europe, especially in England. In many underdeveloped countries, the value of protecting bats has not yet been recognized, and in most, vulnerable species are killed in ignorance or hunted for food.

Despite the major and well-documented decline of bats in the United States, only endangered gray, Indiana, Ozark big-eared, Hawaiian hoary and Virginia big-eared bats are federally protected. Bats are often protected under state game laws, but such laws are rarely enforced where bats are concerned. Real protection is rarely possible without cooperation from an educated public. ∎

Merlin Tuttle, an internationally recognized authority on bats, is curator of mammals at the Milwaukee Public Museum. For more information on protecting bats, write Bat Conservation International, Milwaukee Public Museum, Milwaukee, Wisconsin, 53233.

It is not as well-known as
its red-haired cousin, but the
gray fox may have more
claims to distinction

The Fox That Climbs Trees

BY WILLIAM CURTIS

I wasn't quite prepared for what happened on a chilly morning last fall when I sent several wailing squalls from my predator call reverberating across a brushy canyon in northern California. After several minutes of complete silence, one of the most bloodcurdling screams I ever heard burst forth from just behind my right shoulder. I couldn't choke back a yell of surprise as I twisted around only to see a gray fox eyeballing me from a distance of ten feet, the hairs along its neck and shoulders standing straight up. It didn't seem possible that such a small animal — eight pounds at the most — could have been responsible for such a hair-raising outburst. At my movement, the disillusioned fox quickly disappeared into the brush.

That was only the third time I've heard gray foxes give forth such vocal demonstrations, and I've lived around these animals all of my life. Apparently, they reserve such screams for special occasions, perhaps to vent anger, fear or frustration, or possibly to provoke a reaction from a stranger in their territory. Knowing that I work with the U.S. Fish and Wildlife Service as a predator control field assistant, people have often called upon me to investigate reports of screaming mountain lions. And many times I've found fox tracks in the area but no lion tracks.

The gray fox is one of the most widely distributed carnivores in the 48 states, but many people are not even aware of its existence. Furtive but ingenuous, this canine is surprisingly easy to trap. It is almost completely lacking in the guile of its cousin, the crafty red fox.

The gray exhibits other traits decidedly not in keeping with the public image of foxes. For one, it climbs trees, easily and regularly, the only North American canid to do so. For another, it is surprisingly facile at adapting to man's activities, often living right among unsuspecting human neighbors.

Many outdoorsmen believe the gray fox can ascend only leaning or brushy trees, but nothing could be farther from the truth. One time our neighbors had some stray chickens ambushed by a varmint. The chicken house was near a brushy creek slicing its way down off a nearby ridge and they called on me for help. It was impossible to find tracks, but I did spot some white feathers strewn along the rocky streambed. Then I located a medium-sized blue oak with a pile of white leghorn feathers scattered around its base.

The bark on the trunk was freshly skinned and scratched from frequent

It's the claws that do it. Long, sharp and curved, they enable the gray fox to shinny right up a tall tree for a good look around.

use. Twelve feet above ground level I could see a fair-sized hole in the ramrod-straight trunk. I just knew that the black, decaying hole marked the entrance to a raccoon's den, and old "Blackcheeks" was the culprit carrying away chickens. I leaned a broken limb against the trunk and climbed up.

Just as I began my ascent, a gray fox popped out of the hole and sailed to the ground like a giant flying squirrel. I could plainly hear the whimpering of young animals inside the hollow trunk and reaching the hole, I saw four tiny pups — their eyes still closed — curled up in a nest of wood chips. Though I'd found many fox dens in more normal locations — culverts, rock piles and hollow logs — this was the first time I knew that grays would den in a vertical tree with no low limbs.

The animals climb so well because their toenails are longer, more curved and much sharper than are those of the red fox. The gray fox's tracks resemble a house cat's far more than they do those of the red fox.

Red foxes range from almost white or buffy hues to about every shade of red. But there is little variation in color even among the subspecies of grays. They are uniformly steel-gray on the back with some longer black-tipped hairs. The throat and belly are white; the sides, legs and ears are reddish brown. The gray's tail is comparatively slender, with a black dorsal ridge and tip, compared to the red's bushy, white-tipped tail.

With their predilection for nighttime activity, gray foxes often thrive close to man without anybody knowing they are around. A short while back, I visited some friends living in California's Berkeley Hills, a residential section of the Bay Area literally bristling with houses. One afternoon, I took a walk up a narrow, chaparral-choked draw. Rooftops and television antennas poked up everywhere. Yet, within two hundred yards, I spooked out two gray foxes as I pushed through the dense brush. Tracks, both old and fresh, were all over the damp ground.

Gray foxes are seldom a major threat to domestic fowl, though on occasion they do take free-roaming chickens, ducks and young turkeys. Jackrabbits, cottontails and brush rab-

Frolicking about their den, these pups were conceived in late winter, born in early spring. After just six months at home, they'll be off on their own.

What do they eat? Rabbits are the gray foxs' favorite prey but, they also eat rats, mice and lizards. Some months, in some areas, they eat only berries.

220

bits rate as blue-plate specials on the gray's menu, especially during the winter and spring months. They can't outrun a jackrabbit over a long haul, but they are capable of short bursts of great speed and consequently they can outmaneuver their prey, catching it against steep banks or canyon walls.

Gray foxes aren't exactly finicky feeders. Besides the favored bunnies, they go for wood rats, mice, squirrels, ground-roosting or nesting birds and their eggs, insects, lizards, frogs and even fish if they can rake them out of creek puddles that are drying up. Garbage dumps and carrion are prime drawing cards. Some fruits and berries rank high. Along the Pacific slopes, they'll thrive on an almost straight diet of manzanita berries in the fall and even gobble down bitter juniper berries.

Gene Trapp, a biologist at California State University in Sacramento, reports that he has found green scorpions over three inches long in the scats of gray foxes in southern Utah on a number of occasions. The principal diet of gray foxes, he adds, are grasshoppers and crickets in the summer, prickly pear fruits in the fall and juniper berries and mule deer carrion in spring.

Gray fox populations fluctuate in cycles and the animals appear to be quite susceptible to the diseases of dogs. I once came across two dead foxes at a mountain spring and, cutting them open, found three varieties of parasitic worms balled up in their stomachs and intestines. Rabies and infestations of scabies curb their numbers at times, but distemper may be responsible for the big die-offs.

Some conservationists have become concerned about gray foxes, due to the soaring price of furs. But although they are easy to trap, they have such wide distribution it would appear virtually impossible to trap them down to a critically low population. They are found from coast to coast and border to border except for the northern Rocky Mountains and Great Plains. They range from the high mountains down to the lowest valley floors. Even if the population was severely reduced on the open ranges, foxes would soon overflow from the protected areas. Once the fox numbers dwindled, the expenses of running long lines in order to make only small catches would not prove feasible for the trapper.

Ideal gray fox terrain in California

would probably be high, brushy ridges broken by numerous washes, draws and canyons, grading down to rolling oak or pine foothills. The gray is rarely seen far from cover of some kind. However, it has learned to adapt — and very well — to man's ways and will thrive better in brushier second growth than in virgin timber.

In north central California, where I live, the gray fox cycle peaked about 12 years ago, and for 5 years they were as numerous as I had ever seen them. Then, in 2 short years, the population fell to a rock bottom low from some disease, maybe distemper. Fur prices were low, then, and there was practically no trapping in the hills. During the past 3 years, the animals have started a slow but steady comeback despite the big jump in fur prices and the accompanying increase in trappers.

In many areas of the northern and central states, gray fox numbers have declined somewhat, but only in regions where thousands of acres are cleared annually to produce more grain. Gray foxes avoid open fields, feeling unsafe wherever they lack cover. By contrast, red foxes are animals of the open fields and meadows, which is probably why they are so much better known.

Gray foxes are probably not trackers; instead, they rely on their excellent vision to spot prey. One winter morning, I was driving up a dry creekbed not far from where I live. I spotted a pair of foxes daintily mincing across an opening on a distant hillside. I rolled down the window of my pickup and blasted out three or four squalls on my predator call, imitating the high-pitched scream of an injured rodent. I wanted to see if the calls would reach the grays.

Curiously, one of them continued on its way just as if it had heard nothing. The second fox wheeled around and came charging down through the brush straight for the pickup. There were a few old wrecked cars scattered around and my pickup didn't spook the creature at all. It had pegged the origin of the calls with amazing accuracy, stopping no more than 15 yards in front of me. Just to see how it would react, I blew out a few squeaks. The fox stayed put, flattened out to no more than a few inches in height, literally poised on its toes ready to leap at the first glimpse of the intended prey. Then, the fox

There's no mistaking this gray. Look for the steel-gray back and the black-tipped tail. A red fox, on the other hand, has a white-tipped tail.

Snooping around abandoned farm buildings is one of the gray fox's favorite pastimes. Given half a chance, the critter will move right in.

Huddled in a hollow log is how young gray foxes are often found. They might also be found under barns, beneath occupied chicken houses, in culverts, in rock piles and in slots and crevices along the bases of cliffs.

pushed itself up with its front legs, stretching out as high as it could in a sitting position. It repeated these calisthenics three times, and each time its movements were so slow I had to watch closely to see any motion. Almost everything adult wild creatures do is for utilitarian purposes and I feel certain that this fox was moving itself up and down to change its sighting plane for a more thorough examination of the area from which it thought the wails were originating.

The life expectancy of a gray fox in the wild is short. In a study in the Southeast, one biologist found that 50 percent of a litter dies the first summer, 90 percent dies before the first winter is over and 50 percent die every year after that. Though some foxes will live for many years in the wild, most do not survive.

Gray foxes would have to rank among the most-easily-gentled creatures found in the woods. They seldom fight among themselves or with any other animal. Several times, I've seen dogs grab and kill foxes that were cornered or trapped with no resistance whatsoever.

One fall morning, while riding a high mountain trail, I met a cowboy. He was holding something under his blue denim jumper. We pulled up our horses to visit for a few minutes and he told me how one of his shepherd pooches had started baying at a hollow oak stump. Peering in the hole, the cowboy saw a fully-grown gray fox hunkered against the back of the cavity. The fox showed no inclination to bite, so he simply reached in and pulled it out and was taking the critter home for a pet.

The little gray fox may never take sly red Reynard's evil place in children's stories, and it is far too secretive to be well-known, but it is an interesting brush-lover that often follows streams right down to residential areas and sets up housekeeping within the clamor of major cities. If you are reading this article anywhere in the U.S. besides the northern Rockies or the Great Plains, there may be a gray fox closer than you think. ∎

Californian William Curtis has worked for the U.S. Fish and Wildlife Service for the past 29 years.

The wonderful anti-ant machine

BY EMILY AND PER OLA D'AULAIRE

WITH its elephantine nose and shaggy body as big as a dog's, the giant anteater looks and behaves like no other creature on earth. It shuffles along on its knuckles, its daggerlike claws turned back and tucked into neat little pockets in the palms of its "hands." When it swims, it uses its long tubular snout as a snorkle. It sometimes uses its tail as an umbrella. But its specialty is eating insects. With its acute sense of smell, it can sniff out ants at 40 paces, then gobble 30,000 in a single day. Its two-foot-long tongue flicks in and out as often as 160 times a minute.

Few beasts are so instantly recognizable. Yet for all its bizarre appendages and unlikely habits, this wonderful anti-ant machine — like the world's other two types of anteater — remains something of a mystery. In fact, the anteater is so secretive in the wild that it's seldom seen there.

Despite all this specialization, the anteater is actually a primitive mammal. Along with the armadillos and sloths, it is one of the last survivors of a primitive order called *Edentata*. There are four species of anteater: the giant version, which weighs more than 100 pounds; two species of lesser anteater, or tamanduas, which are shorter, with more monkeylike tails; and the silky or pygmy form, which weighs but a pound or so. All live in the tropical forests or savannahs of Central America and South America; like all wildlife there, they are plagued by rampant development and habitat destruction.

Latin America's habitat woes have made captive breeding programs more important for preserving the gene pools of unique creatures such as anteaters — and anteaters have turned out to be entirely different beasts in captivity than they are in the wild. At the Leningrad Zoo, for example, a breeding pair of giant anteaters thrived on an amazing diet. Each one ate the following daily ration: ten mice and a half-pound of minnows, a pint of milk, a pound of hamburger, some mashed carrots and onions, a pinch of chalk, a touch of hedge rose syrup and — for old time's sake — 1¾ ounces of Grade

A ant pupae. No one has ever seen wild anteaters eating anything except an assortment of insects.

Studies of captive animals have helped unravel some of the mysteries about the anteater's sense of smell. The creature's long snout is filled with olfactory bulbs, but no one knew how efficiently an anteater could sniff out ants until two biologists trained a captive animal to associate the pungent odor of camphor with food. The animal could detect camphor even when 4,000 times as much eucalyptus was used to mask its smell. By comparison, human subjects couldn't detect camphor when only 100 times as much eucalyptus was used to hide its odor. Thus, an anteater can easily find its insect prey from a considerable distance. And the animal has to rely heavily on this amazing sense of smell, because it's so nearsighted that a human standing five feet away is nearly invisible.

Another fascinating thing about the anteater is its circulatory system. Dennis Meritt, Assistant Director of Chicago's Lincoln Park Zoo, says that the animal is partially cold-blooded: "Because of an unstable body temperature, anteaters are much more dependent on their surroundings to keep their bodies warm enough to function. They get in trouble very quickly if they become chilled." To prevent this from happening, anteaters utilize the principle of heat exchange. The arteries and veins in the legs and tail are interlaced in such a way as to bring the cooler venous blood close to the hotter arterial flow, thereby functioning as a heat pump and air conditioner.

The anteater's strange, shuffling gait derives from yet another internal peculiarity — double-jointed hips. The shuffle is most obvious in the giant anteater, which resembles a distorted Afghan hound as it moves across the savannahs and rain forest floors.

The giant is the most familiar anteater species because of its size and terrestrial habits. It measures up to nine feet from end to end, and stands two feet high at the shoulder. It is also the most formidable of the three forms. On each front foot, it has three razor-sharp claws that are up to four inches

They may look like Rube Goldberg contraptions, but anteaters, from the lesser shown here to the giant, are quite adept at using their outlandish noses, knifelike claws and flypaper tongues to seek out and eat hordes of insects in a single day.

While the giant anteater prowls the ground in search of food, the mysterious one-pound pygmy never leaves the treetops

long, keen as switchblades and capable of eviscerating attacking jaguars with a single swipe. The giant anteater leads a nomadic existence, curling up for naps wherever it happens to be when sleepy. Sometimes it digs depressions in the soil where it lies down, head between forelegs and tail pulled over it like a comforter. In this position, it's all but invisible against the ground. Its gray-brown hair is stiff and coarse and its coat has a distinctive white stripe.

The tail is a brushlike affair that streams out like a great banner. But at the front end, it's hard to tell just where the neck leaves off and the head begins. The body simply tapers to a blunt point that's equipped with tiny nostrils and mouth a foot or so beyond the small, bleary eyes.

One of the oddest facets of the giant anteater's behavior is its mode of raising its young. Breeding takes place once a year, though if the baby should die, the mother may mate again. After a six-month gestation period, the female gives birth to a single baby weighing about 17 pounds. After two or three weeks, the baby grabs hold of the mother's hair, scampers nimbly up her tail and settles on her back where it clings tightly in cozy contentment. Exactly like its mother in all but size, it will ride this way until it is a year old and almost as large as the parent, a double-decker effect of weird beast upon weird beast that must have had early explorers to Latin America won-

dering if they had partaken too heavily of fermented manioc.

Apparently, giant anteaters — or all four species, for that matter — have idiosyncrasies about gathering their insect food. Some adopt the bull-in-a-china-shop approach, knocking open any anthill or termite mound they find. Then, despite painful bites from the aroused prey, in goes the tongue, twisting and turning, to the very center of the colony. When enough ants are trapped on the tongue's sticky surface, the appendage is rapidly withdrawn. The process is repeated until the nest is empty, or the anteater apparently can tolerate no more bites around the face, ears, nose and mouth. And then there are those perhaps apocryphal anteaters who let the ants — or any of some 30 different species of bugs — do most of the work. As reported by one seventeenth-century English observer, an individual may lie down, snout flat on the ground, and extend its flypaper tongue languidly across an ant trail like a pale red welcome mat. In two or three minutes, he wrote in his journal, the tongue would be covered with ants. All that was left for the creature to do was to pull in the mat and swallow.

Whereas the giant anteater spends most of its time on the ground, its cousins, the two species of tamandua (Tupi Indian for "catcher of ants"), forage for insects in the trees as well as on the forest floor. A tamandua's fur is shorter than that of a giant anteater,

Off goes the top and in goes the tongue — at 160 strokes a minute — as a giant anteater raids a termite mound. The predator has an acute sense of smell and can sniff out termites from as far away as 40 paces.

A baby (right) rides on its mother's back across Brazil's savannah.

Francisco Erize

F. Gohier (Photo Researchers)

Among the most bizarre mammals on earth, anteaters face a double threat: people shoot them and destroy their living space

and its tail is bare on the underside and at the end. The tail can be twisted around branches for support as the tamandua breaks open tree insect nests with its claws. When threatened, this dog-sized animal, like other anteater species, rears up on its hind legs and spreads its arms wide in a come-and-get-me challenge, displaying two-inch-long awl-like claws. As with the giant anteater, the tamandua's young ride on the mother's back, but such rides take place mainly at night. The animal usually spends the daylight hours curled up, almost invisibly, high in the treetops or in holes in the ground. Although the tamandua is easily approached, scientists know less about it than they do about its larger relatives.

But the most mysterious of the four species is the smallest, the silky or pygmy anteater — so named for its size and its soft golden fur. Tiny enough to hold in one hand, a tenth the size of the tamandua, this one-pound creature rarely leaves the treetops and seldom moves about during the day. Unlike the other species, this anteater's nose is stubby, and its eyes are large for night vision. The tail is highly prehensile for treetop wandering.

Despite these differences, though, the silky anteater is unmistakably an anteater. Family life is similar, and so is defense against enemies — those sharp claws again, which even in this diminutive beast are not to be trifled with. "Like the other anteaters, once the silkies get ahold of something, it's almost impossible to get them to release their grasp," declares Dennis Meritt, who has studied them in captivity. "Once they grab you, it's like being caught in a miniature vise. The nails are so sharp they go right through thick leather gloves — and I have the scars to prove it."

So does John Walsh, author of *Time Is Short and the Water Rises*, the account of a program to save wild animals from the waters behind a new dam in Surinam. Walsh tells of holding a newly rescued silky in his hand and thinking how charmingly the soft, furry ball uncurled, stood on its hind legs, wrapped its tail around one of Walsh's fingers and put its paws alongside its nose, prayer-fashion. Then suddenly, the animal bowed forward and dug sharp claws deep into Walsh's wrist. With a howl, and a new respect for anteaters, whatever their size, he dropped the creature into the boat.

One of Walsh's assistants learned a more severe lesson from a tamandua. The animal was clinging to an abandoned termite nest a few feet above the water when the rescuers found it. As the men tried to pry it loose, it tumbled into the reservoir. Instinctively, the assistant reached down to snatch the animal to safety. The creature wrapped its prehensile tail around his wrist and gave him a hooked wallop that tore a chunk out of his forearm. And a giant anteater can be an even meaner customer. When cornered, this animal has reportedly killed more than one person with its powerful forelegs and sharp claws.

Although anteaters are important to humans because they help control crawling insects, these unique and bizarre mammals face an uncertain future. Habitat destruction is the most prevalent danger: the felling of rain forests plus cultivation and grazing of the savannahs where the animals feed.

The giant anteater is in the most trouble because it is the largest and least likely to flee from humans. It is totally protected by law in Brazil, and its meat is all but inedible — yet it is often shot on sight. Sometimes, its fur is used in the tourist trade, and articles made of anteater hair, including belts, watch fobs and tote bags, turn up in local markets. The giant also runs afoul of the cattle industry in such nations as Venezuela and Colombia. Says Dennis Meritt, "Granted, anteaters don't eat the same things as cattle and they shouldn't be viewed as competitive. But whenever man is involved — well, the situation changes." But one thing is certain: If a person had to choose between anteaters and swarms of insects, he'd take anteaters, every time. ∎

Emily and Per Ola d'Aulaire have profiled a menagerie of creatures — from vultures to lemmings — for magazines ranging from Reader's Digest *to* Scanorama.

Giant anteaters are not shy around humans, and many are shot on sight. In this instance, the rider means no harm.

Francisco Erize

Some scientists are skeptical of reports that anteaters, when cornered, have killed people — but this tamandua looks mean enough to do the job after being surprised by the photographer along a rain forest road. The truth is, these beasts are normally docile and have more to fear from people, who are rapidly destroying anteater habitat.

Nanook Bears Watching

And that's exactly what scientists are doing, as they try to uncover the mysteries of polar bears

STORY AND PHOTOGRAPHS
BY FRED BRUEMMER

WE were drifting north through the Bering Strait, prisoners of the ice. My Eskimo companions seemed unconcerned. They were rugged subsistence hunters from Little Diomede, that bleak, craggy island located halfway between Alaska and Siberia. I had joined them to observe a remote group of walruses. When the vast, shifting fields of pack ice compacted and entrapped us, the men hauled our skin-covered boat onto a large ice floe, and there we waited for conditions to improve. We dozed off to the sounds of the pack: the creak and groan of the grinding floes; the shrill cries of the ivory gulls; the distant roar of the walrus herds that were drifting north, like us, upon the ice.

"Nanook!" one of the Eskimos suddenly exclaimed, shaking me from my sleep. He pointed to a large male polar bear that was shuffling over the ice in that peculiar, rolling pigeon-toed gait that caused nineteenth century whalers to call the creature "the farmer."

His coal black nose twitching, the bear stopped momentarily and sniffed at the air. Apparently, he had picked up the scent of a pod of walruses. He ambled toward the animals and, at 100 yards, he charged — a yellowish blur across the glittering ice. The alarmed marine mammals poured like a brown avalanche off their floe into the sea. Polar bears rarely attack adult walruses, but scientists believe that they may intentionally spook a herd, hoping to isolate a helpless calf. The ploy didn't work this time, though. The bear turned and ran off to the west into what was, for us, the forbidden territory of the Soviet Union.

Polar bears are, of course, not bound by human borders, and their realm is immense: five million square miles of circumpolar land and frozen sea. They have been sighted in what one early explorer, Peter Freuchen, called "the most horrible part of the world: the ice desert near the North Pole." And they have been found as far south as Canada's James Bay, which is on the same

Shuffling across the ice in its distinctive rolling gait, a polar bear appears clumsy, but when pressed, it is surprisingly fast and agile. In recent years, scientists have revised some of their long-held beliefs about the animal Eskimos call "Nanook."

231

Hungry bears devour the remains of a seal, their principal prey in the Arctic. At times, a polar bear will crouch for hours beside a seal's breathing hole, waiting for the marine mammal to surface.

latitude as London, England. In all, the bears range through the territory of five nations — the United States, Canada, Norway, Denmark and the Soviet Union — yet we have only recently begun to understand their movements.

Over the past 20 years, scientists have captured, tagged and studied more than 3,000 polar bears. In doing so, the researchers have revised some of their long-held beliefs about the animals. Some biologists now think that polar bear populations are much more stable than was once presumed.

Ever since 1296 when the Venetian traveler Marco Polo reported that "bears of white color and prodigious size" existed far to the north in "the Region of Darkness," naturalists believed that the creatures were eternal wanderers, passing at random from one continent to the next. Even the Eskimos who inhabited the bears' domain shared that belief. Recent tagging and radio-tracking studies, however, have found that most of the bears live within relatively constrained geographical areas; most of the study animals never

wandered more than 100 miles from the place where they were captured.

Today, many researchers think that the bears form distinct populations within six specific regions: Alaska, Central Siberia, the Canadian Arctic Archipelago, Greenland, and Spitzbergen and Franz Josef Land (two areas north of Norway). Apparently, bears of different populations also differ in size. The smallest of the animals are found in the populations north of Norway. The largest are those that live off the coast of western and northern Alaska. There, some males may stand 11 feet and weigh more than three-quarters of a ton. Along with Kodiak bears, they are the largest land carnivores on earth.

In 1961, the Russian biologist S.M. Uspenskij estimated that only between 5,000 and 8,000 polar bears existed throughout the world. American scientists were more optimistic, putting the figure at about 12,000 animals. But everyone agreed that the bears needed protection. The Soviet Union had already banned hunting of the creatures in 1956. The United States followed suit with a partial ban in the early 1970s. Finally, in 1973, the five "polar bear nations" signed an agreement to protect both the animals and their habitat.

Since then, the gloomy picture has

brightened. Most contemporary estimates, which are still subject to disagreement among experts, now put the bears' worldwide population at a minimum of 20,000. Even under the most favorable conditions, though, polar bears increase their numbers slowly. Females do not mate until they are four or five years old. Then, they have one or, more often, two cubs at three-year intervals. An individual female will usually not bear more than seven or eight cubs during her lifetime. Thus, polar bear numbers can recover from natural or man-caused declines very slowly.

The bears mate from April through June, and by November, most pregnant females retreat to special denning areas that the creatures have used for untold generations. On Wrangell Island north of Siberia, for instance, Uspenskij found about 150 such maternity denning sites. On the coast of Alaska, however, no concentrated denning areas have been discovered; females den in widely scattered areas where conditions are suitable. Some polar bears even den far from land on the ice of Beaufort Sea.

The den is usually an oval chamber five to ten feet long, five feet wide and about three feet high. It is excavated in a snowbank, often on the leeward side of a ridge or riverbank. Since snow is

A Canadian scientist tags a tranquilized bear to study the creature's movements. Researchers have found that the gigantic predators live within relatively limited geographical areas; most of the animals studied stayed within a 100-mile radius.

an excellent insulator, the temperature within the den may be 40 degrees F. higher than that outside. The cubs, usually born in December, weigh only about 1½ pounds and have a thin coat. They cannot hear, and their eyes do not open for 40 days. They snuggle into their mother's fur and drink her fat-rich milk.

While the females with newborn cubs winter in a snug den, the other bears roam the ice in darkness and in bitter cold. When it storms, they sleep, either in temporary dens or out in the open. I once watched several polar bears snooze soundly through a 48-hour blizzard. The snow drifted against the animals and eventually covered them. When the storm ended, the bears rose, yawned, stretched, shook the snow out of their shaggy fur and ambled on.

A thick layer of insulating fat and extremely dense, oily, water-repellent wool, covered by a coat of long guard hairs, enables polar bears to endure the worst arctic weather. Recent studies using scanning electron microscopes reveal that the animals' hairs are transparent and hollow, and have reflective inner surfaces. Researchers believe that the hairs can transmit solar energy to the polar bears' black skin. In 1978, officials at the San Diego Zoo were as-

tounded when some of their polar bears turned green. Algae were thriving in the bears' "greenhouse" coats.

In March, females and young emerge from their dens into the dazzling glitter of the arctic. The cubs, now densely furred, weigh a chubby 20 pounds. They often stay with their mother for 2½ years. At first, they nurse and share her meals. Then they start to hunt — awkwardly at first, but later with increasing skill. Together, the females and cubs head for the ice where they find their principal prey, ringed seals — the most common and widespread of all arctic pinnipeds.

Polar bears use three main methods to catch seals: the patient wait, the stealthy stalk and the "surprise" attack. Hunkered down near a seal's breathing hole in the ice, a bear often crouches for hours. When the seal finally surfaces, the waiting bear pulls it onto the ice with powerful jaws.

In late spring and early summer, seals bask upon the ice, and the bears stalk them with caution and finesse. When the seal sleeps, the bear advances. The instant the seal wakes and looks around, the bear becomes an immobile lump upon the ice. At about 30 yards, the bear charges and grabs its prey. Meanwhile, on pack ice where the seals rest

at the edge of floes, a polar bear may ease itself into the water and swim just beneath the surface. Then, at the floe's edge, the animal surges out of the water and pounces upon the startled seal.

The polar bear has always been important to Eskimos. They prize its superb pelt and, until fairly recently, held special ceremonies to appease the creature's spirit. But even with "spiritual" assistance, the Eskimos' hunting success was relatively low. On rough ice, pleated with pressure ridges, the animal could easily outdistance a dog team and sled. Despite its lumbering appearance, a polar bear, when pressed, is amazingly fast and agile (a young male can attain a speed of about 35 mph). I once watched a bear that was being pursued scale a 35-foot ice wall with seemingly effortless ease. Between 1925 and 1945, when polar bears were hunted primarily by natives in Alaska, the average annual kill was only about 125 bears.

In the late 1940s, however, trophy hunting of the bears (usually with airplanes used to locate the animals for

Their frigid domain has been called "the most horrible part of the world," but polar bears can withstand even the worst climatic conditions. A thick layer of fat, covered by a coat of long, hollow guard hairs, insulates them against cold.

people on the ground) increased in Alaska. The number of bears taken there gradually increased to 300 a year; the total worldwide kill of polar bears reached 1,300 in 1964 and rose to 1,500 in 1967. Many experts agreed that the animals could not sustain predominately unregulated hunting indefinitely. "They were hitting those bears pretty hard," recalls Charles Jonkel, a bear biologist at the University of Montana. Jonkel and other researchers noted that polar bears were no longer being seen onshore in areas where they had once been common. Their concern precipitated, in 1965, the first meeting of the world's top polar bear experts, which was held in Fairbanks, Alaska. These scientists have since met biennially to coordinate research and exchange information.

Because the bears live in such remote regions, it is difficult to estimate their populations accurately. The Alaskan population is usually estimated at between 6,000 and 9,000 bears, and is generally thought to be stable. "I always try to emphasize that the populations appear to be stable *at very low levels*," says Steve Amstrup, Polar Bear Project Leader for the U.S. Fish and Wildlife Service. "Any threat to the security of the bears could be disastrous." Habitat alteration and unregulated harvests are considered the leading threats to that security.

That day a few years ago, as my companions from Little Diomede and I watched the animal from our icebound boat, I thought about how superbly adapted the polar bear is to its harsh northern realm. The destruction of critical portions of that realm by energy exploration and other development could create mammoth new problems for the creatures in the years ahead. Nevertheless, at a recent meeting, the world's polar bear experts were cautiously optimistic. Prospects for the bears look good, they said, "if mankind is careful." ◼

Canadian Fred Bruemmer is a writer-photographer who specializes in northern regions. He spends part of each year living with native people.

THE DAM BUILDER IS AT IT AGAIN!

Scientists are putting beavers to work on several ravaged western streams, and the results are turning skeptics into true believers

By Phillip Johnson

DRIVING west from the town of Lolo into Montana's Bitterroot Mountains, biologist Greg Munther slows his dusty beige pickup beside a stretch of Howard Creek. To the untutored eye, the deep channel cut through a grassy meadow looks like the postcard image of a Montana valley. But to Munther, 40, the steeply banked creek, bordered by willows trimmed into lollipop shapes by cattle chewing at their lower branches, is a portrait of a stream in trouble.

Howard Creek's overgrazed banks are eroding, and the stream has deepened into a gully. With fewer and fewer plants to hold moisture in the valley's soil, the groundwater level is dropping. If the sad cycle that has already hit thousands of western valleys plays out here, the creek will slow to a trickle during the summer months, losing its quicksilver crop of rainbow and cutthroat trout. As the stream cuts deeper, the valley's surface soil will lose moisture. Acres of succulent shrubs and green grasses could well become a parched range, all

237

Building a better home, beavers dammed this creek in Alaska (left) and put up a lodge to house their family. In doing so, they also created new habitat for fish and game. Researchers believe the beaver's engineering prowess can help combat erosion and other serious problems.

LEONARD LEE RUE IV

All in a day's work: a beaver drags a branch back to its lodge (below). During the warmer months, some of the animals may gnaw down 200 trees or more. Over the centuries, such constant construction work has had an impact on the face of almost every watershed in North America.

but barren for most cattle and wildlife.

Munther has found a way to break through that cycle in Howard Creek and elsewhere in North America—a simple, inexpensive technique for protecting healthy streams and revitalizing desolate valleys. His method: restoring to streams one of their chief architects over millions of years. Along with a small but growing band of resource managers and scientists, Greg Munther has rediscovered the continent's largest rodent, *Castor canadensis*, the North American beaver.

Like people, beavers shape their environment to survive. By damming creeks and digging channels, the slow-moving, virtually defenseless vegetarians give themselves escape routes and sheltered lodges. But their dams also create soft, mucky banks rather than eroding creek edges. The dammed streams run high enough to moisten arid lands, creating wet wilderness valleys that welcome caddis flies and coho salmon, muskrats and moose, ospreys and redstarts. Grazing animals (including cattle) skirt the edges of such streams, finding forage in moist meadows without trampling down the banks. That's why Munther and other wildlife managers are now turning to the beaver to remedy such man-made problems as erosion and overgrazing, lack of water and loss of fish. What beavers provide to streams and entire valleys is productivity.

"I was seeing all these devastated streams," recalls Munther, a U.S. Forest Service employee and avid sportsman. "But then I started noticing that streams that had beavers were in fine shape, even with a lot of grazing nearby, while others that had lost their beavers were decimated. It was as if a light bulb went on in my head. I thought, 'There's something there.'"

Munther began to hire trappers to catch beavers with gentle Bailey live-traps in settled areas such as the Clark Fork River basin. Then he and his crew released the beavers on streams like Howard Creek, where, he says, they "waddled down to the stream, sniffed the water and waded right in."

If they are left alone here, he adds, "I'm pretty sure that this valley will just be a string of ponds." And that in turn will provide more pools for trout to rear young in, more cover for fish and birds, more mud for insects, more rotting organic matter for microbes, more forage

for browsing animals—in all, a richer community of plants and animals, in the water and around it.

To prove his point, Munther drives across a ridge to the Fish Creek drainage. Here fresh beaver sign is plentiful. A large beaver dam creates a deep pond around a jumbled pile of willow branches, which is the beaver's lodge. Munther steps onto the dam, and a sudden gust of sulphurous fumes arises from the decomposing muck, mingled with the odor of crushed mint. Like many of nature's most fecund neighborhoods—swamps, salt marshes, tidal flats—beaver streams are not conventionally pretty. But they teem with life. "This is one of the best fishing streams we've got on the Lolo Forest," says Munther. "The beavers improve the pools, make them places for fish to survive the winter." In Fish Creek, what the beavers have created is habitat, for everything from algae to eagles.

The beaver had a different value to the earliest European explorers in North America. For two centuries beginning in the early 1600s, the word "beaver" was synonymous with "hat" in fashionable Europe. Beaver pelts lured Europeans to most of North America's wild reaches; the beaver trade was the main cause of the French and Indian War. Between 1853 and 1877, the Hudson's Bay Company alone shipped three million pelts to Europe. But the nineteenth century invention of machines that could produce top-quality felt led to a change in fashion.

By then, beavers were a devastated species: there had been as many as 400 million beavers in North America before the Europeans came, but by 1890, only small populations survived, predominately in the Midwest. Today, the animals have made a slow comeback: there are now thought to be more than two million in the United States.

Unfortunately, civilization hasn't left the creatures much room. Conflicts between beavers and people today concern culverts blocked by beaver dams, or flooded hay fields, or stands of trees gnawed down at night. When their instinctive drive to impound running water runs afoul of civilization, the creatures are labeled a "nuisance" and destroyed or deported.

Ecologists call the beaver a "broad-niched" animal. The creatures are found a few miles north of the Arctic Circle and in the Sonoran Desert of Mexico. Once they ranged throughout Europe and Siberia; today the population outside North America has been reduced to remote areas in Scandinavia and the U.S.S.R. (except for a few survivors that continue to hang tough in the Rhone River). They can live at altitudes approaching 12,000 feet and below sea level.

In the craggy deserts of southwestern Wyoming, Bruce Smith, 34, a fisheries biologist for the U.S. Bureau of Land Management (BLM), began to appreciate the beaver when, like Greg Munther, he grew concerned after being assigned to the agency's Rock Springs district. Smith learned that less than one percent of the terrain was still "riparian"—influenced by surface water. He discovered that 83 percent of the wetlands that once existed in his territory were so badly degraded, chiefly by intensive grazing, that their productivity has lessened. Only two percent of the area's streambanks were in something resembling their "natural" state.

These cold desert regions, now empty and forbidding, looked very different to pioneers and Indians in the 1800s. What are now dry washes were once wet meadows where wagon trains pastured stock. Then the river valleys were greener and more inviting. Streams that now run thick and brown through stark, slumped-in canyons once meandered through ponds and marshes, supporting stands of aspen, cottonwood and willow, and teeming with trout. Valleys chosen by early settlers have long since been abandoned as uninhabitable. What is missing from the current picture is the beaver.

Smith first began to realize this when field researchers discovered the last surviving remnants of two subspecies of trout, the Colorado and Bear River cutthroat. Most such native subspecies have been overwhelmed by hatchery and imported fish, but it was learned that the genuine articles still live in the headwaters of two remote streams near Cokeville, Wyoming. They were in such waters, isolated and protected by dense networks of beaver lodges and dams. Recalls Smith: "We began to see the beaver ponds as crucial to a healthy stream."

Smith's early efforts to encourage beavers as a way of restoring habitat met with a misfortune that also plagued Greg Munther—someone killed the beavers. So along with Larry Apple, a transplanted Iowan who serves as a BLM biologist, Smith tried again on two less-traveled watersheds, Sage and Currant creeks.

Sage Creek is a dramatic example of the erosional cycle that afflicts the region. Historically, it had been occupied by beavers, but heavy grazing pressure had cut down tree growth, and the beavers were eventually forced out. Once their dams and the surrounding vegetation were gone, the stream rapidly carved a deep wound in the earth. Through this gash as much as 200 tons of silt a day bled downstream into Flaming Gorge Reservoir.

Remarkably, when Smith and Apple arrived, a few beavers were trying to recolonize Sage Creek. Another ecologist's epithet for the beaver is "choosy generalist"; the beaver prefers aspen, eating the bark and using the wood for dam construction, and after that it will choose willow, but if it has to, it can make use of just about anything. In Sage Creek, with nothing else to choose from, the beavers were using sagebrush.

But sagebrush made fragile dams. High flows would wash them out, leaving the beavers unprotected and the streambanks bare. Smith had an idea: why not give the beavers something to work with? Assisted by the Boy Scout troop he was scoutmastering at the time, Smith tried cutting some aspen high up on the mountain slopes and trucking it down to the gully. "They went wild," he says. "As long as we kept bringing the aspen, they kept building dams." Then he and Larry Apple had another inspiration. They scrounged a load of used truck tires, carted them to Sage Creek and distributed them over a small dam. The beavers, generalists in no position to be choosy, promptly adapted to this new environmental circumstance and incorporated the tires into a larger and more stable dam. Sage Creek was ultimately to sport five truck-tire dams, covering a half-mile stretch of the stream.

The results were striking. Silt moving down the creek was blocked by the dams. During the course of a single winter, that of 1981-82, the ponds silted in—the main pond, 60 feet long and 36 feet wide, was solid mud in two weeks.

This raised the streambed and the surrounding water table several feet, and created shelves of mud on which sedges, bulrushes and willows took root. The spreading water began to drown the sagebrush and greasewood, making room for grasses and forbs. Deer and many species of birds reinhabited the area. It is too early and too muddy for fish to return to Sage Creek, but on Currant Creek, which was in better shape, the trout population exploded as fish migrated from Flaming Gorge.

"We keep stressing that we're using beaver for habitat recovery," Smith emphasizes. "We're not really managing for beaver." He estimates that the effort has cost about $4,000 for materials, plus the time of federal employees. In contrast, erosion control structures built by human engineers for these extreme conditions could cost upwards of $100,000.

In certain American Indian creation stories, it was said that great beavers fetched the mud with which the Earth was built. Modern science tells a somewhat similar tale.

The beaver's first known direct ancestor, *Steneofiber*, appeared nearly 40 million years ago. A creature like the modern beaver has been living in North America for at least one million years. Until the end of the last Ice Age, the beaver shared North America with a giant cousin, *Castoroides*, which measured nearly ten feet from nose to tail and may have weighed 600 pounds.

At one time or another during all these eons, beavers have probably modified almost every watershed on the continent. Modern beavers are building marshes and meadows atop the foundations laid by thousands of their predecessors. For that matter, the gullies that beaverless streams cut today usually erode through soils that were deposited behind the dams of past centuries.

The beaver made an especially dramatic contribution to northerly landscapes, recolonizing ice-gouged valleys after the glaciers retreated. Time after time, geologists or archeologists have found that the first layer of organic material lying above glacial deposits is an ancient beaver pond, with twigs and branches that show the marks of the beavers' teeth. One case in point: a 1960 excavation down to bedrock revealed that the area now called Boston Common was created by a beaver marsh.

Throughout North America, the beaver played a dominant role in the laying-down of topsoil in the fertile valleys. If the beaver didn't build the Earth, it did fetch the mud for many of the tillable portions of the northern hemisphere. In a sense, we are living today on income from the beaver's capital. It would be well to remember this when the animal's needs conflict with human designs. The farmer who begrudges the beaver a few flooded acres may well owe the very soil that he farms to the beaver's efforts in the distant past. Often the fisherman who resents a dam that temporarily blocks migrating trout forgets the beaver's role in keeping the stream productive.

What is both exciting and sad about the programs started by Greg Munther and Bruce Smith is that both are "pioneering" in territory that has been discovered many times before. Programs to repropagate the rodents seem to arise every other decade. Forty years ago, the federal government actually parachuted the animals into wilderness areas of Idaho, in special traps that opened on impact. They were also transplanted to Washington, Utah, Wyoming and Oregon (the Beaver State had lost most of its beavers by the time it achieved statehood). Harold Ickes, Secretary of the Interior under Franklin Roosevelt, boasted that he got $300 worth of conservation from beavers that cost $5 per head.

Current research, though, has led scientists to a new understanding of the beaver. Jim Sedell and Cliff Dahm regard it as a "keystone species"—one that plays a pivotal role in regulating an entire ecosystem.

Sedall and Dahm, both Oregon natives, are a team based at Oregon State University; Sedell, 39, is a U.S. Forest Service stream ecologist stationed at the school; Dahm, 32, is a researcher in the university's department of fisheries and wildlife. Like Munther and Smith, they were led to the beaver by studying fish. While investigating the decline of coho salmon, Sedell and Dahm discovered that the fish depended on the stored water and increased productivity of beaver ponds. Typically, a few stretches of a creek influenced by beaver activity produce a large proportion of the river system's young fish.

Cliff Dahm studies nature at the most basic level. One-cell bacteria and diatoms are the beginning of the food

Wading into a Montana stream, Greg Munther (below) is one of several U.S. scientists who consider the beaver an important ally. Like Munther, plaid-shirted Bruce Smith and associate Larry Apple, about to release a beaver into a Wyoming creek (bottom), "discovered" the big rodent while studying fish.

Investigating the decline of coho salmon in Oregon, ecologist Jim Sedell (taking a water sample, below) learned that the fish depend on the stored water and increased productivity in beaver ponds. Sedell and his associates think that beavers may ultimately increase the fertility of entire river systems.

PHILLIP JOHNSON

PHILLIP JOHNSON

ROBERT CHILDS (BLM)

Young beavers are born in the spring and remain with their parents through the entire year. By the second autumn, they leave and build their own lodges.

chain. From Dahm's point of view, it might be said that the most important thing a beaver does is to provide the right chemistry for such microscopic organisms to grow.

Beaver ponds trap sediment drifting down the stream. They also generate a great deal of debris themselves, nourishing plants which grow, die and decay. Beavers drag a tremendous amount of wood and bark into the stream. This organic material sinks to the bottom of the pond and decomposes. Seething decomposition sucks every bit of oxygen from the muck; because it lacks oxygen, scientists call the brew "anaerobic."

Two vital things take place in such a brew. First, a great deal of carbon and nitrogen is recycled so one-celled organisms in the water can use it. In the coastal streams of the Northwest, nitrogen is generally in shortest supply. The biochemistry of beaver ponds makes nitrogen available to more organisms, increasing a stream's capacity to support everything from microbes to mammals. (In eastern streams, phosphorus is more likely to be the scarce nutrient, and beaver ponds boost its availability as well.)

Secondly, in the anaerobic mud at the bottom of the pond, specialized bacteria ingest many metals and other minerals that might otherwise have washed down to the sea. When they make their way out of the muck, the minerals have been transformed into suitable nutrients for other forms of bacteria and algae. The rich populations of microbes in a beaver pond are the beginnings of a food web leading up to salmon, herons and bears. Says Cliff Dahm: "They stimulate the system at every level."

This new notion of the workings of a stream, with the beaver as the keystone, might change some conventional ideas about the way organisms relate to each other, which is what the science of ecology is all about. "I look at it as an alternative to the traditional concept of the pyramid-shaped food web," notes Dahm. "In the pyramid, you start with your lowest-level feeders, which are fed upon, and finally you end up at the top with fish or mammals. The idea has al-

ways been that you're feeding up to the top. Here there is something at the top—'the beaver'—which we feel has a major impact on all the lower levels. We're kind of inverting all the arrows."

The nutrients released in beaver ponds seep through the dams and travel downstream. Beavers do more than enhance the wildlife habitat in their immediate vicinity. They increase the fertility of entire river systems.

In prehistoric North America, it was not uncommon to find animals shuttling resources around in dramatic fashion. "You had flocks of billions of birds and herds of millions of animals," says Jim Sedell. "All their trampling and pooping had a tremendous impact on the ecosystem. Passenger pigeons' influence on phosphorus and nitrogen cycles must have been ferocious. And certainly there's been nothing to replace the bison as a big, mobile form of fertilizer. The same thing with the salmon, when there were millions of them dying in the rivers every year. They were keystones." They all add nutrients to the system in more usable forms. Today, only the tenacious beaver continues to play so important a role in shaping the environment.

There is one more ecologist's word which helps to explain the relationship between people and beavers: "sympatric." Two organisms are sympatric when they share the same habitat. Our difficulty is that we are heavily sympatric with the beaver. Thanks to the animal's work in stimulating fertility and shaping the landscape, the beaver's habitat has been highly productive for human beings. So far, this has been to the detriment of beavers, but, in the long run, loss of the beaver reduces the productivity of watersheds for all their denizens, including *Homo sapiens*.

If beavers were simply engineers of dams and waterways, we could simulate their role, although at tremendous cost. But perhaps they are something else entirely, a keystone that cannot be removed without losing the arch. As late-coming cohabitants of America's streams and river valleys, we humans may yet need to learn how to live in a sympatric relationship with *Castor canadensis*. ∎

Oregon writer Phillip Johnson spent several weeks last summer traveling through the American West to interview scientists in preparation for this article.

GROWING BY LEAPS

In a mountainous world, young goats must learn survival in a hurry

STRANGE visions sometimes come to those who venture onto the mountaintops. Close to the crest of the Great Divide in Montana, I once peered over a rocky ledge and saw a mountain goat kid riding on its mother's back as she trotted along. After a moment of confusion, I realized that the youngster must have been clambering across its resting nanny when she suddenly stood up to begin moving.

I shouldn't have been surprised to see such fancy footwork by a young mountain goat. In the seven years I studied these alpinists in Montana, I saw many demonstrations of their extraordinary climbing abilities. This mountain rodeo was just one more example—not only of the powerful inborn climbing tendencies of baby goats, but also of the remarkably close relationship which exists between the nannies and their offspring. These two factors are undoubtedly central to the survival of the youngsters, which grow up in a niche higher and steeper than that of any other large species in North America.

The American mountain goat is native to the peaks of the continent's northern Pacific coast as well as to the northern Rockies. The name goat is a bit misleading: this peculiar, bearded, dagger-horned beast is actually the New World's sole representative of the small rupicaprid, or goat-antelope, tribe. Its other members are Europe's chamois and Asia's lesser-known goral and serow. Originally found only from southern Alaska to Washington, Idaho and Montana, mountain goat populations have been transplanted during this century to several dozen new ranges in

Springing into a dangerous world, a mountain goat kid leaps across a crag in Washington's Olympic Mountains. Kids can perform such feats soon after birth.

DAVID C. FRITTS

244

AND BOUNDS

By Douglas Chadwick

the West. The animals now range as far south as Nevada, Utah and Colorado.

Wherever they find themselves these days, mountain goats remain closely bound to cloudscraping cliff habitats throughout the year, living at altitudes up to 10,000 feet. In Spring, the young goats are born. Toward the last half of May, pregnant nannies begin to withdraw from herds and seek out steeper-than-average sections of the cliffs. There, away from the distractions of goat society, and protected from predators, they rest and wait. In a day or two, each will have delivered a single, snow-flake-white infant weighing about the same as a human baby—six to eight pounds—into a precarious existence.

Immensely valuable climbing skills, and the fearlessness that goes with them, seem to be born into mountain goats. I watched one kid come into the world under some Douglas firs one year during the last week of May. Within hours, it was climbing its resting nanny, then clambering over a medium-sized boulder. Without hesitation, the baby sprang from there high into the air— only to land more on its nose than its feet. Despite its flexible hooves and traction pad soles, the kid's wobbly legs could not match all its inborn aspirations. Yet it went back up the boulder and sprang again.

This seeming recklessness may be a necessary survival skill in the risky mountain world, but it still puts kids in a variety of hair-raising situations. That is why an unusually close nanny-kid relationship is also necessary; the nanny must be nearby at all times for protection. Her main chore is to help her enthusiastic but still rubbery-legged kid avoid injuring itself or spilling over the

A baby goat (above) often uses its nanny, resting after an arduous pregnancy, for climbing practice before it moves on to bigger, steeper outdoor surfaces.

Twins are relatively uncommon among mountain goats. And chances are, one of these kids (right) will not survive its first high-country winter.

cliff edge on a one-way trip. Accompanying her young, step by teetering step, the nanny generally stations herself close below, with her body between the infant and the risky side of things.

Hours-old kids not only climb as if their future depended on it, but exhibit basic versions of adult fighting behavior—whirling to hook their head at imaginary opponents—and sexual postures. They quickly outgrow "baby food," too; on the second day of life the infant begins sniffing and mouthing the types of plants eaten by its mother. By the fourth day, already nursing much less often, it grazes alongside her for short periods.

Before the first week is over, the nanny grows increasingly restless. She is ready to rejoin the herd, and the kid is now ready to keep up—almost, that is. Although the nanny's role as a living safety net is no longer as important, her work as a climbing guide takes on new meaning once the pair start traversing new terrain. Whenever the kid becomes stuck along a precipitous route, or merely hesitates, the nanny rushes back to its side; sometimes to urge it forward, but more often to pick out a less rough-and-tumble path.

The steep environment has fewer big predators than the lowlands—one reason why mountain goats stay where they are. However, I've seen golden eagles

*Nannies stay within a few feet of their
kids at all times, warding off hawks and cougars
and keeping the youngsters from cliff edges*

occasionally try to snatch kids. A stamping, snorting, 140-pound adult nanny brandishing two wickedly pointed stilettos provides an impressive defense as its kid, tail upraised in fear, huddles beneath her. Cougars, large and agile on rocky terrains, may be the most serious four-legged threat to the goats in many ranges, according to Washington game biologist Rolf Johnson.

The young goat is not even safe when it joins its relatives. The white climbers have a prickly social temperament to match their sharp rupicaprid horn and are quick to defend their personal space against other herd members. Before I began watching goats, I had read about their propensity for fighting, but nothing prepared me for what I saw. In one eight-member band of goats that I observed my first year in the Swan Mountains in Montana, some animals were involved in nearly two dozen fights in a single hour — fights that sometimes end with puncture wounds from the sharp horns. Young goats are not excluded; I watched one kid in the same band being butted off a 15-foot drop by an adult nanny.

It is understandable, then, that a nanny remains aggressively protective of her kid as the pair join any of the various small bands that make up a herd. The kid, in turn, will most often be found within a yard of two of her side throughout the entire 10 to 11 months of their association, even though it is essentially weaned after one month.

As the warm days continue, the newest additions to the herd feed across melt-water-moist pastures and frolic together in mock combat under the

A herd enjoys a mineral lick during springtime (left). Highly flexible hooves keep mountain goats secure on slopes approaching a 90-degree pitch.

Mountain goats grow up in some of the highest altitudes in North America (top, right), grazing above the timber line where there are few dangerous predators.

Curious and sociable, baby goats (right) will turn their playful explorations of the world into the knowledge that keeps them alive when they leave their nannies.

Storms and predators will take their toll;
only half of the goats in a typical herd
will survive their first mountain winter

DAVID C. FRITTS

watchful eye of their nannies. But then, often as soon as the first of October, fresh snow arrives to reclaim the high country. Abruptly, the ledges become glazed with ice, Forage is buried beneath drifts and crusts. And all the while, the potential for avalanches, a significant cause of mountain goat deaths, keeps building.

Winter may be the single biggest threat facing a young goat, and the climbing skills it has developed, together with the protection of its nanny, are called to the test. A single storm can drop a pile of flakes chest-high to a 225-pound adult billy, which means ear-high to a 40-pound kid. The tougher the conditions, the more the youngster must rely upon its nanny to break trail and dig down to expose food—and to defend those hard-earned feeding sites from others. Lacking the strength and climbing skills of older goats, juveniles struggle harder to get from place to

Goat society involves a constant struggle for dominance, and youngsters are often not spared. This yearling (above) is being pushed by an older goat.

After years of evading cougars, wading through six-month winters and fighting other goats, an adult billy (right) has earned his rest on a summer snowpatch.

snowy place. Lacking, too, the fat reserves and simple heat-retaining bulk of bigger goats, the young are at the mercy of cold extremes and must use up still more precious energy keeping warm.

By the time a new green season begins, only half the kids in a typical herd will be around to welcome it. Weighing perhaps 50 to 75 pounds, if that, the previous generation, now yearlings, is also subject to heavy overwinter losses. The small advantage they possess in size and strength, compared to kids, is partially offset by the fact that they no longer enjoy maternal help and protec-

tion. Like two-year-olds, they must settle for following different adult females at a respectful wary distance. They will not mature sexually until age two-and-a-half or reach full size until age four.

This is the bargain the mountain goat has struck with its environment: in return for a home free from many of the usual pressures of predators and competition with other hooved animals, it must meet all the tests of winter, towering rocks and hostile society at the upper edge of the life zone—and meet those tests from the moment of birth. By the time a kid reaches adulthood each mountain goat will have won the right to produce its own batch of feisty, precocious kids. ∎

Douglas Chadwick lives in Polebridge, Montana, on the edge of Glacier National Park. His most recent book, A Beast the Color of Winter *(Sierra Club Books, 1983), is about his studies of mountain goats.*

CREDITS

Articles are listed in the order in which they appear in the book. Except where noted, all articles are copyrighted by the National Wildlife Federation. The copyright date is the date of publication.

Feathers by David Cavagnaro, *National Wildlife,* June-July, 1983, pp. 20-25, ©1983 David Cavagnaro.

Fairy Tales are True by Lisa Yount, *National Wildlife,* August-September, 1983, pp. 46-51.

Diary of a Loon Watcher by Jen and Des Bartlett, *International Wildlife,* January-February, 1981, pp. 40-48.

Storks Get a Helping Hand by Nigel Sitwell, *International Wildlife,* November-December, 1984, pp. 14-19.

Can Filipinos Learn to Love this Bird? by Robert S. Kennedy, *International Wildlife,* July-August, 1983, pp. 26-33, ©1983 Robert S. Kennedy.

Shore Leave, *National Wildlife,* February-March, 1985, pp. 4-11.

I Focus on Kenya's Birds by M.P. Kahl, *International Wildlife,* May-June, 1976, pp. 20-27, ©1976 M.P. Kahl.

Splendor at the Shore, *National Wildlife,* June-July, 1977, pp. 42-47.

Cool in California by Vic Cox, *National Wildlife,* February-March, 1984, pp. 52-53.

Big Red Booms in Cincinnati by Steve Maslowski, *National Wildlife,* December-January, 1983, pp. 12-17.

This Raptor Keeps a Low Profile by Frances Hamerstrom, *National Wildlife,* August-September, 1977, pp. 42-47.

Meet the Real Nesters of Australia by Michael Morcombe, *International Wildlife,* May-June, 1973, pp. 20-27, ©1973 Michael Morcombe.

Birds of a Different Bill, *National Wildlife,* August-September, 1984, pp. 4-5.

Blueprint for Bluebirds by Eileen Lanzafama, *National Wildlife,* April-May, 1984, pp. 24-28.

He Migrates with the Seasons by Nyoka Hawkins, *National Wildlife,* October-November, 1983, pp. 46-53.

Something to Gobble About by Renira Pachuta, *National Wildlife,* October-November, 1982, pp. 24-28.

Hornbill by Paul Spencer Wachtel, *International Wildlife,* January-February, 1982, pp. 36-41.

Homing in on the Hunter by David Seideman, *National Wildlife,* April-May, 1984, pp. 50-55.

Courting with Gold, *International Wildlife,* March-April, 1985, pp. 12-13.

Body Snatcher by Norman Myers, *International Wildlife,* May-June, 1983, pp. 36-43

Dancing To A Different Beat by Tui De Roy Moore, *International Wildlife,* January-February, 1983, pp. 44-47.

Ornament and Armament by Stephen Freligh, *International Wildlife,* May-June, 1984, pp. 42-47.

Charging Through the Seasons by Renira Pachuta, *National Wildlife,* June-July, 1981, pp. 40-47.

Monkey Business by Edward R. Ricciuti, *International Wildlife,* November-December, 1983, pp. 16-23.

Desert Machines, *International Wildlife,* November-December, 1982, pp. 4-13.

Hail, Hail, the Gang's All Here by Mark Wexler, *National Wildlife,* October-November, 1979, pp. 24-31.

Taking Off the Velvet by George H. Harrison, *National Wildlife,* October-November, 1981, pp. 46-47.

A School for Elephants by Denis D. Gray, *International Wildlife,* January-February, 1978, pp. 36-43.

Phantoms of the Prairie by Michael Tennesen, *National Wildlife,* June-July, 1976, pp. 4-11.

The Small Cats of Africa by Norman Myers, *International Wildlife,* March-April, 1976, pp. 4-11, ©1976 Norman Myers.

You Can't Sneak up on a Marmot by Jerome J. Knap, *International Wildlife,* May-June, 1977, pp. 42-47, ©1977 Jerome J. Knap.

In Celebration of Bats by Merlin D. Tuttle, *International Wildlife,* July-August, 1983, pp. 4-13, ©1983 Merlin D. Tuttle.

The Fox that Climbs Trees by William Curtis, *National Wildlife,* June-July, 1977, pp. 20-27.

The Wonderful Anti-ant Machine, by Emily and Per Ola D'Aulaire, *International Wildlife,* November-December, 1979, pp. 24-29.

Nanook Bears Watching by Fred Bruemmer, *National Wildlife,* December-January, 1983, pp. 37-43.

The Dam Builder is at it Again by Phillip Johnson, *National Wildlife,* June-July, 1984, pp. 8-15.

Growing by Leaps and Bounds by Douglas Chadwick, *National Wildlife,* August-September, 1984, pp. 44-51.

PHOTO CREDITS

Cover: Owl eyes by Robert Carr; Tiger eyes by Dana Bodnar
Page 2: Flamingo by Frans Lanting; Serval cat by Norman Myers (Bruce Coleman, Inc.).
Pages 4-5: Fred Bruemmer
Pages 6-7: Entheos
Page 9: Robert Carr
Page 133: Dana Bodnar

Cover Design by Tim Kenney

Library of Congress
Cataloging-in-
Publication Data

Birds and mammals.

1. Birds. 2. Mammals. I. National
Wildlife Federation.
QL673.B56   1986   598   86-12508
ISBN 0-912186-73-9

NATIONAL WILDLIFE FEDERATION
1412 Sixteenth Street, N.W., Washington, D.C. 20036-2266